T3-BNK-452

NCRP REPORT No. 116

LIMITATION OF EXPOSURE TO IONIZING RADIATION

**Recommendations of the
NATIONAL COUNCIL ON RADIATION
PROTECTION AND MEASUREMENTS**

Issued March 31, 1993

National Council on Radiation Protection and Measurements
7910 Woodmont Avenue   /   Bethesda, MD 20814

LEGAL NOTICE

This Report was prepared by the National Council on Radiation Protection and Measurements (NCRP). The Council strives to provide accurate, complete and useful information in its reports. However, neither the NCRP, the members of NCRP, other persons contributing to or assisting in the preparation of this Report, nor any person acting on the behalf of any of these parties: (a) makes any warranty or representation, express or implied, with respect to the accuracy, completeness or usefulness of the information contained in this Report, or that the use of any information, method or process disclosed in this Report may not infringe on privately owned rights; or (b) assumes any liability with respect to the use of, or for damages resulting from the use of any information, method or process disclosed in this Report, *under the Civil Rights Act of 1964, Section 701 et seq. as amended 42 U.S.C. Section 2000e et seq. (Title VII) or any other statutory or common law theory governing liability.*

Library of Congress Cataloging-in-Publication Data

National Council on Radiation Protection and Measurements.
    Limitation of exposure to ionizing radiation : recommendations of the National Council on Radiation Protection and Measurements.
      p.   cm. — (NCRP report ; no. 116)
    "Issued April 1, 1993."
    Includes bibliographical references and index.
    ISBN 0-929600-30-4
    1. Ionizing radiation--Safety measures. I. Title. II. Series.
RA569.N353   1993
612'.01448--dc20

                                      93-7142
                                       CIP

Preface

This Report updates and replaces National Council on Radiation Protection and Measurements (NCRP) Report No. 91, *Recommendations on Limits for Exposure to Ionizing Radiation.* Although the recommendations contained in this Report are similar to those in NCRP Report No. 91, the Council desires to reiterate and update its position on radiation protection issues following the publication of additional data on the biological effects of ionizing radiation by the National Academy of Sciences/National Research Council Committee on the Biological Effects of Ionizing Radiations (BEIR V), the United Nations Scientific Committee on the Effects of Atomic Radiation, and the review of these documents by Scientific Committee 1-2 of the NCRP that is being published as NCRP Report No. 115, *Risk Estimates for Radiation Protection Purposes* and the publication of the *1990 Recommendations of the International Commission on Radiological Protection.* Deviation in the recommendations of this Report from those of the ICRP reflect the Council's desire to incorporate greater flexibility or increased protection in its recommendations for those situations where it is reasonable to do so.

Serving on NCRP Scientific Committee 1 for the preparation of this Report were:

Charles B. Meinhold, *Chairman*
National Council on Radiation
Protection and Measurements
Bethesda, Maryland

Members

Seymour Abrahamson	**S. James Adelstein**
University of Wisconsin	Harvard Medical School
Madison, Wisconsin	Boston, Massachusetts

William J. Bair
Battelle Pacific
Northwest Laboratories
Richland, Washington

R.J. Michael Fry
Oak Ridge National
Laboratory
Oak Ridge, Tennessee

John D. Boice, Jr.
National Cancer Institute
Bethesda, Maryland

Eric J. Hall
Columbia University
New York, New York

Edward W. Webster
Massachusetts General Hospital
Boston, Massachusetts

Advisor

Warren K. Sinclair
National Council on Radiation
Protection and Measurements
Bethesda, Maryland

NCRP Secretariat

William M. Beckner

The Council wishes to express its appreciation to the Committee members for the time and effort devoted to the preparation of this Report.

Charles B. Meinhold
President, NCRP

Bethesda, Maryland
15 March 1993

Contents

1. Introduction

The National Council on Radiation Protection and Measurements (NCRP) published its last complete set of basic recommendations specifying dose limits for exposure to ionizing radiation in NCRP Report No. 91 which was published in 1987 (NCRP, 1987). During the preparation of that report, three factors were recognized as important consequences of the emerging information from the continuing study of the atomic bomb survivors by the Radiation Effects Research Foundation (RERF). The first was the continued appearance of excess cancers observed during the latest survey period. Second, these cancers were appearing at a rate consistent with a multiplicative projection model. The third factor was the effect on risk estimates of revised dose estimates. These factors all suggested that there would be increases in projected risk. However, since the anticipated new risk estimates were unavailable, the Council employed the risk estimates given by the International Commission on Radiological Protection (ICRP) in its Publication 26 (ICRP, 1977). In Report No. 91, the NCRP recommended an annual occupational dose limit of 50 mSv and an annual limit for members of the public (excluding natural background and medical exposures) of 1 mSv for continuous exposures and 5 mSv for infrequent annual exposures. At that time, however, the Council anticipated a potential increase in risk estimates. Consequently, it encouraged a control on lifetime occupational exposure and cautioned the user to consider the dose limits as upper limits rather than design goals.

Now that the United Nations Scientific Committee on the Effects of Atomic Radiation (UNSCEAR, 1988), the National Academy of Sciences/National Research Council Committee on the Biological Effects of Ionizing Radiations (BEIR V) (NAS/NRC, 1990), the International Commission on Radiological Protection (ICRP, 1991a) and NCRP Scientific Committee 1-2 (NCRP, 1993a) have completed their risk assessment activities, the Council has reexamined its 1987 recommendations. This Report is the result of this reexamination and it replaces in its entirety NCRP Report No. 91, *Recommendations on*

1

Limits for Exposure to Ionizing Radiation (NCRP, 1987). The basic framework of this Report, the approach to dose limitation and the principle of a Negligible Individual Dose (NID), however, are based on the earlier report (NCRP, 1987).

The recommendations and concepts provided in ICRP Publication 60 (ICRP, 1991a) have been carefully reviewed and in the interest of a uniform international approach to radiation protection have, in general, been incorporated in this Report. Deviation from their recommendations was deemed necessary in a few cases where greater flexibility could be obtained at similar or less risk (*e.g.*, the occupational dose limits) or where increased protection was considered to be warranted (*e.g.*, a monthly exposure limit for the embryo-fetus). Table 1.1 provides a comparison of the radiation risk data, recommendations, and other factors used in NCRP Report No. 91 (NCRP, 1987) and ICRP Publication 60 (ICRP, 1991a) with those used in this Report.

TABLE 1.1 — *Comparison of NCRP Report No. 91 (NCRP, 1987) and ICRP Publication 60 (ICRP, 1991a) with this Report.*

Recommendations, Risk values and Other factors	NCRP Report No. 91 (NCRP, 1987)	ICRP Publication 60 (ICRP, 1991a)	This Report
Assumed Radiation Risks			
Workers	1.25×10^{-2} Sv^{-1} for fatal cancer[a]	4.0×10^{-2} Sv^{-1} for fatal cancer	4.0×10^{-2} Sv^{-1} for fatal cancer
		0.8×10^{-2} Sv^{-1} nonfatal cancer detriment	0.8×10^{-2} Sv^{-1} nonfatal cancer detriment
	0.4×10^{-2} Sv^{-1} for severe genetic effects[a]	0.8×10^{-2} Sv^{-1} for severe genetic effects	0.8×10^{-2} Sv^{-1} for severe genetic effects
Members of the Public	(not specifically addressed)	5.0×10^{-2} Sv^{-1} for fatal cancer	5.0×10^{-2} Sv^{-1} for fatal cancer
		1.0×10^{-2} Sv^{-1} for nonfatal cancer	1.0×10^{-2} Sv^{-1} for nonfatal cancer
		1.3×10^{-2} Sv^{-1} for severe genetic effects	1.3×10^{-2} Sv^{-1} for severe genetic effects
Embryo-fetus	20×10^{-2} Sv^{-1} total detriment (UNSCEAR, 1986)	(not specifically stated)	$\sim 10 \times 10^{-2}$ Sv^{-1}

TABLE 1.1 — *continued*

Recommendations, Risk values and Other factors	NCRP Report No. 91 (NCRP, 1987)	ICRP Publication 60 (ICRP, 1991a)	This Report
Occupational Dose Limits[b]			
Based on Stochastic Effects	50 mSv annual effective dose equivalent limit *and* 10 mSv × age (y) cumulative effective dose equivalent guidance for the workplace[c]	50 mSv annual effective dose limit *and* 100 mSv in 5 y cumulative effective dose limit[c]	50 mSv annual effective dose limit *and* 10 mSv × age (y) cumulative effective dose limit[c]
Based on Deterministic Effects	150 mSv annual dose equivalent limit to lens of eye *and* 500 mSv annual dose equivalent limit to all other organs[d] (*e.g.*, red bone marrow, breast, lung, gonads, skin and extremities)	150 mSv equivalent dose limit to lens of eye *and* 500 mSv annual equivalent dose limit to skin, hands and feet[d]	150 mSv annual equivalent dose limit to lens of eye *and* 500 mSv annual equivalent dose limit to skin, hands and feet[d]
Annual Limits of Intake (ALI)	$\dfrac{50 \text{ mSv}}{H_E (50) \text{ Bq}^{-1}}$	$\dfrac{20 \text{ mSv}}{E (50) \text{ Bq}^{-1}}$	—
Annual Reference Levels of Intake (ARLI)	—	—	$\dfrac{20 \text{ mSv}}{E (50) \text{ Bq}^{-1}}$

TABLE 1.1 — *continued*

Recommendations, Risk values and Other factors	NCRP Report No. 91 (NCRP, 1987)	ICRP Publication 60 (ICRP, 1991a)	This Report
Public Dose Limits[b]			
Based on Stochastic Effects	1 mSv annual effective dose equivalent limit for continuous exposure *and* 5 mSv annual effective dose equivalent limit for infrequent exposure[c]	1 mSv annual effective dose limit and, if needed, higher values provided that the annual average over 5 y does not exceed 1 mSv[c]	1 mSv annual effective dose limit for continuous exposure *and* 5 mSv annual effective dose limit for infrequent exposure[c]
Based on Deterministic Effects	50 mSv annual dose equivalent limit to lens of eye, skin and extremities[d]	15 mSv annual equivalent dose limit to lens of eye *and* 50 mSv annual equivalent dose limit to skin, hands and feet[d]	50 mSv annual equivalent dose limit to lens of eye, skin and extremities[d]
Embryo-fetus	5 mSv dose equivalent limit and a dose equivalent limit in a month of 0.5 mSv once pregnancy is known[d]	2 mSv equivalent dose to the woman's abdomen once pregnancy has been declared and limiting intakes of radionuclides to about 1/20 of an ALI[d]	0.5 mSv equivalent dose limit in a month once pregnancy is known[d]
Negligible Individual Dose (NID)	0.01 mSv annual effective dose equivalent per source or practice[e]	—	0.01 mSv annual effective dose per source or practice[e]

TABLE 1.1 — continued

Recommendations, Risk values and Other factors	NCRP Report No. 91 (NCRP, 1987)	ICRP Publication 60 (ICRP, 1991a)	This Report
Organ or Tissue Weighting Factor (w_T)[a]			
Gonads	0.25	0.20	0.20
Red bone marrow	0.12	0.12	0.12
Colon	—	0.12	0.12
Lung	0.12	0.12	0.12
Stomach	—	0.12	0.12
Bladder	—	0.05	0.05
Breast	0.15	0.05	0.05
Liver	—	0.05	0.05
Esophagus	—	0.05	0.05
Thyroid	0.03	0.05	0.05
Skin	0.01	0.01	0.01
Bone surface	0.03	0.01	0.01
Remainder	0.30	0.05	0.05

TABLE 1.1 — *continued*

Recommendations, Risk values and Other factors	NCRP Report No. 91 (NCRP, 1987) $\bar{Q}$	ICRP Publication 60 (ICRP, 1991a) w_R	This Report w_R
Radiation Weighting Factor (w_R) and Quality Factor ($\bar{Q}$)			
X and γ rays, electrons, positrons and muons	1	1	1
Neutrons			
Thermal neutrons	5		
Neutrons other than thermal	20		
Energy < 10 keV		5	5
10 keV to 100 keV		10	10
> 100 keV to 2 MeV		20	20
> 2 MeV to 20 MeV		10	10
> 20 MeV		5	5
Protons			
Energy > 2 MeV		5	2
Alpha particles, fission fragments, nonrelativistic heavy nuclei	20	20	20

[a]In NCRP Report No. 91 (NCRP, 1987), it was recognized that the total risk estimate of 1.65×10^{-2} Sv^{-1} for a working population and possibly the values for the organ or tissue weighting factor (w_T) would change as a result of the reassessment of the data for the Japanese survivors that was then underway.

[b]The dose limits exclude medical and natural background exposures.

[c]The concepts of effective dose equivalent and effective dose are different (see Section 5).

[d]The concepts of equivalent dose and dose equivalent are also different (see Section 4).

[e]In this Report, the Negligible Individual Risk Level (NIRL) introduced in NCRP Report No. 91 (NCRP, 1987) is changed to a Negligible Individual Dose (NID) with the same numerical value of 0.01 mSv but without a corresponding risk level (see Section 17).

2. Radiation Protection Goals and Philosophy

2.1 Goal of Radiation Protection

The goal of radiation protection is to prevent the occurrence of serious radiation-induced conditions (acute and chronic deterministic effects) in exposed persons and to reduce stochastic effects in exposed persons to a degree that is acceptable in relation to the benefits to the individual and to society from the activities that generate such exposures.

2.2 Effects of Concern in Radiation Protection

The serious radiation-induced effects of concern in radiation protection fall into two general categories: deterministic effects and stochastic effects.

A deterministic effect is defined as a somatic effect which increases in severity with increasing radiation dose above a threshold dose. The severity increases because of damage to an increasing number of cells. Deterministic effects occur only after relatively large doses, but the threshold dose and the severity of the effects are influenced by individual susceptibility and other factors. The effects may be early, occurring within hours or days; or late, occurring months or years after exposure. Examples of acute or early effects are erythema and other skin damage. Chronic or late effects include lens opacification that may lead to impaired vision, loss of parenchymal cells, fibrosis, organ atrophy and a decrease in germ cells that may result in sterility or a reduction in fertility.

The question of radiation dose thresholds for deterministic effects is complex and the magnitude of the apparent threshold depends on the specific biological endpoint and the ability to detect it. However, if the endpoints of concern are restricted to those that are clinically

8

significant, dose limits can be selected to be less than the threshold values for these effects.

Certain clinically significant deterministic effects of radiation exposure of the embryo-fetus, *i.e.*, structural anomalies or abnormal development or growth, may have low dose thresholds during gestational periods that are highly critical for organogenesis. Such effects may increase in frequency with absorbed dose and may also have the deterministic character of increasing severity with absorbed dose. In humans, development and growth of the central nervous system is particularly radiosensitive in this regard over specific periods of time during gestation (see Section 10).

For the purpose of this Report, a stochastic effect is defined as one in which the probability of the effect occurring increases continuously with increasing absorbed dose while the severity of the effect, in affected individuals, is independent of the magnitude of the absorbed dose.[1] A stochastic effect is an all-or-none response; for example, the occurrence of cancer. There are differences in the risk of an effect for a given dose that are dependent on individual factors such as age, sex, etc. A stochastic effect might arise as a result of radiation injury in a single cell or in a substructure such as a gene and is assumed to have no dose threshold, although currently available observations in population samples do not exclude zero effects at very low doses. The induction of stochastic effects (cancers and genetic effects) is considered to be the principal effect that may occur following exposure to low doses of ionizing radiation.

2.3 Objectives of Radiation Protection

The specific objectives of radiation protection are:
(1) to prevent the occurrence of clinically significant radiation-induced deterministic effects by adhering to dose limits that are below the apparent threshold levels and
(2) to limit the risk of stochastic effects, cancer and genetic effects, to a reasonable level in relation to societal needs, values, benefits gained and economic factors.

---

[1]This is assumed to be true for humans at those absorbed doses not involving other effects such as cell killing that may predominate at higher absorbed doses.

These objectives can be achieved by ensuring that all exposures are As Low As Reasonably Achievable (ALARA) in relation to benefits to be obtained and by applying dose limits for controlling occupational and general public exposures.

Based on the hypothesis that genetic effects and some cancers may result from damage to a single cell, the Council assumes that, *for radiation-protection purposes, the risk of stochastic effects is proportional to dose without threshold, throughout the range of dose and dose rates of importance in routine radiation protection.* Furthermore, the probability of response (risk) is assumed, for radiation-protection purposes, to accumulate linearly with dose. At higher doses, received acutely, such as in accidents, more complex (nonlinear) dose–risk relationships may apply.

Given the above assumptions, radiation exposure at any selected dose limit will, by definition, have an associated level of risk. For this reason, NCRP reiterates its previous recommendations (NCRP, 1987) concerning:

(1) the need to justify any activity which involves radiation exposure on the basis that the expected benefits to society exceed the overall societal cost **(justification)**,

(2) the need to ensure that the total societal detriment from such justifiable activities or practices is maintained **ALARA**, economic and social factors being taken into account and

(3) the need to apply individual dose limits to ensure that the procedures of justification and ALARA do not result in individuals or groups of individuals exceeding levels of acceptable risk **(limitation)**.

Optimization, as defined by the ICRP in its Publication 37 (ICRP, 1983) and Publication 55 (ICRP, 1989a) is recognized to have the same meaning as ALARA, which is the term that will be used in this Report.

This Report is primarily concerned with the second and third principles specified above, namely, ALARA and dose limits. As will be seen in Section 8, the dose limit is the upper limit of acceptability rather than a design criterion. For example, it is inappropriate to design a barrier based on criteria that would allow individuals to be exposed to the annual dose limit.

In many applications, ALARA is simply the continuation of good radiation-protection programs and practices which traditionally have been effective in keeping the average and individual exposures for monitored workers well below the limits (NCRP, 1987). Approaches employing quantitative estimates of total radiation detriment and costs of protection have been developed by the ICRP (1983; 1989a). Application of these and other quantitative approaches to the making of decisions for maintaining radiation levels ALARA have been presented in NCRP Report No. 107, *Implementation of the Principle of As Low As Reasonably Achievable (ALARA) for Medical and Dental Personnel* (NCRP, 1990a) and are being considered by NCRP Scientific Committee 46-9 on ALARA at Nuclear Plants.

3. Basis for Occupational Dose Limits

3.1 Introduction

Occupational and nonoccupational dose limits have changed over the years in step with evolving information about the biological effects of radiation and with changes in the conceptual framework within which recommended dose limits are developed and applied.

In the 1930s, the concept of a tolerance dose was used, a dose to which workers could be exposed continuously without any evident deleterious acute or early effects such as erythema of the skin.

By the early 1950s, the emphasis had shifted to chronic or late effects. The maximum permissible dose then employed was designed to ensure that the probability of the occurrence of injuries was so low that the risk would be readily acceptable to the average individual (NCRP, 1954). In that decade, based on the results of genetic studies in *drosophila* and mice, the occupational limit was substantially reduced and a public limit introduced. Subsequently, the genetic risks were found to be smaller and cancer risks larger than were thought at the time. The philosophy of the ICRP, as set out in its Publication 26 (ICRP, 1977), and that of the NCRP, as in its Report No. 91 (NCRP, 1987), then came to be concerned, as far as occupational exposure was concerned, with a comparison of the probability of radiation-induced cancer mortality with annual accidental mortality in safe industries. This brief preamble is intended to acquaint the reader with the necessary concept that exposure limits are based on a mixture of observed effects and judgment.

3.2 Comparison with other Industries

The philosophy of NCRP, as established in this Report, is that for occupational exposure, the level of protection provided should ensure that potential stochastic effects are maintained ALARA, commensurate with social and economic factors but, in any case, the risk to an individual of a fatal cancer from exposure to radiation should be no greater than that of fatal accidents in safe industries.[2] It must be recognized that inherent in this decision are arbitrary choices and many uncertainties.

Important elements in the approach that need to be recognized include:

(1) Uncertainty in the risk per unit dose for exposure at high dose and high-dose rate. This is estimated to be uncertain to about a factor of two.

(2) Uncertainty in the extrapolation of risks from exposures at high dose, in the dose region where excess stochastic effects have been observed in humans, to exposures at low dose and low dose rate. This uncertainty is estimated to be about an additional factor of two or more since, at very low doses, the possibility that there is no risk cannot be excluded. This uncertainty is in addition to inherent experimental errors in the data and is most likely the predominant uncertainty in the estimate of risk at low doses.

(3) The choice of the fatal accident risk in safe industry as a measure of acceptability. Many nominally safe industries have annual fatal accident rates of 10^{-4} or less. However, these industries may have substantial morbidity from nonfatal injuries and work-related diseases (ICRP, 1985).

(4) In addition to the decrease in fatal accident rates with time, the detection and treatment of cancer is changing, thereby making the results of such comparisons time dependent.

In this Report, the Council has elected to select dose limits based on a comparison of the fatal accident risk in safe industries with the assumed risk of radiation-induced fatal cancers, a fraction of the nonfatal cancers and severe genetic effects.

Table 3.1 lists the fatal accident rates in the United States for various industries for 1976 and 1991. This Table also shows that the fatal accident rates in the various industries are decreasing with time at the rate of nearly three percent per year.

---

[2]It is recognized that radiation exposure is not the only risk to an individual employed in an industry that involves radiation exposure.

Since NCRP recommends that the level of radiation protection should result in risks that are comparable to or less than those in safe industries, the radiation-protection system should result in an average annual risk of fatal cancer of the order of 10^{-4} or less. Since the dose limit is the maximum permissible dose to be received by a worker, it is reasonable that the risk associated with it be several times higher than this average value, that is, a risk comparable to that of workers in the more hazardous jobs within safe industries, *i.e.*, a fatal accident risk of between 10^{-4} and 10^{-3} per y.

TABLE 3.1 — *Fatal accident rates in various industries, 1976 and 1991.*

	Mean rate 1976[a] (10^{-4} y^{-1})	Mean rate 1991[b] (10^{-4} y^{-1})
All groups	1.42	0.90
Trade	0.64	0.40
Manufacture	0.89	0.40
Service	0.86	0.40
Government	1.11	0.90
Transport and public utilities	3.13	2.20
Construction	5.68	3.10
Mines and quarries	6.25	4.30
Agriculture (1973-80)	5.41	4.40

[a]Reference NSC (1977).
[b]Reference NSC (1992).

It should be kept in mind that there is an inherent limitation to comparing actual deaths occurring among workers in various industries with predicted cancer deaths among radiation workers. Nonetheless, the NCRP average annual occupational dose limit (10 mSv y^{-1}) based on these considerations would result in a cumulative risk over a lifetime from each year's exposure of between 10^{-4} and 10^{-3}. The risk to the *average* worker from such a dose limit should result in a lifetime risk from each year's exposure of one-fourth to one-sixth of this value, *i.e.*, between 2×10^{-5} and 2×10^{-4}.

As part of its multiattribute approach, the ICRP (1991a) focused on the maximum risk that workers have been found to tolerate, *i.e.*, an annual fatal accident risk of 10^{-3} rather than the risk of fatal accidents in safe industries used by the NCRP. However, the risk associated with a lifetime of exposure at the occupational dose limit of either ICRP or NCRP is that due to approximately 1.0 Sv or 0.7 Sv, respectively (see Appendix A).

4. Absorbed Dose, Equivalent Dose and Radiation Weighting Factor

4.1 Introduction

Radiations differ in their relative biological effectiveness (RBE) per unit of absorbed dose. The Council now accounts for this difference by use of the equivalent dose ($H_{T,R}$) which is the product of the average-absorbed dose ($D_{T,R}$) in a tissue or organ (T) due to radiation (R) and a radiation weighting factor (w_R) for each radiation in question,[3]

$$H_{T,R} = w_R D_{T,R}. \qquad (4.1)$$

The radiation weighting factor (w_R) is a dimensionless factor selected to account for the differences in the biological effectiveness of different types of radiation, within the range of doses of concern in radiation-protection activities. This radiation weighting factor (w_R) is specifically related to the type and energy of the incident radiation or, in the case of internal emitters, the radiation emitted by the source. It should be noted that for a tissue or organ, the **equivalent dose** (H_T) is conceptually different from the **dose equivalent** (H). The dose equivalent (H) is based on the absorbed dose at a "point" in

---

[3]$H_{T,R}$ and w_R were first introduced by the ICRP (1991a).

16

tissue which is weighted by a distribution of quality factors (Q) which are related to the LET distribution of the radiation at that point. The equivalent dose, on the other hand, is based on an average absorbed dose in the tissue or organ (D_T) and weighted by the radiation weighting factor (w_R) for the radiation(s) impinging on the body or, in the case of internal emitters, the radiation emitted by the source.

When the radiation field is composed of several types and energies of radiations, *i.e.*, radiations with different values of w_R, the equivalent dose in a tissue (H_T) is the summation of all the incremental, average tissue doses due to each of the component radiations, multiplied by their respective w_R values,

$$H_T = \sum_R w_R \, D_{T,R}.$$

(4.2)

4.2 Basis for the Recommended Values of Radiation Weighting Factor

Although the Council is able to use human data in establishing its risk estimates, such data do not exist for the selection of w_R values. Table 4.1 provides a summary of the limiting RBE_M values for fission neutrons versus gamma rays for a number of biological endpoints.

This data has been used to develop a formal mathematical relationship between the quality factor (Q) and lineal energy (y) or linear energy transfer (L) (ICRU, 1986). At this time, the Council continues to recommend the use of this approach for measurement purposes. For example, the ambient and individual dose equivalent are metrological quantities which incorporate this relationship. The recommended values for Q as a function of L are given in Table 4.2.

As can be seen in Table 4.2, the Q values for low-LET radiation (x rays, gamma rays and electrons) are all designated as unity, even though, at low doses, differences in RBE_M between them have been identified. This is done because the introduction of different quality factors (or w_R values) for different photon or electron energies would suggest a greater reliance on the actuality of these differences than is justified at this time and would, therefore, lead to unjustified

TABLE 4.1 — *Summary of estimated RBE_M values for fission neutron versus gamma rays.*[a]
(Adapted from Table 9.1 of NCRP, 1990b.)

End point	Range of values
Chromosome aberrations, human lymphocytes in culture	34 - 53
Oncogenic transformation	3 - 80[b]
Specific locus mutations in mice	5 - 70[c]
Mutation endpoints in plant systems	2 - 100
Life shortening in mice	10 - 46
Tumor induction in mice	16 - 59

[a]RBE_M is the limiting RBE or the RBE at minimum dose.
[b]The value of 80 was derived from one set of experiments only.
[c]The value of 70, derived from data on specific locus mutations in mice, is not necessarily an RBE_M.

TABLE 4.2 — *Quality factor–LET relationships.*
(Adapted from Table A-1 of ICRP, 1991a.)

Unrestricted linear energy Transfer, L_∞, in water (keV μm^{-1})	$Q(L_\infty)$[a]
< 10	1
10 to 100	$0.32L_\infty - 2.2$[b]
> 100	$300(L_\infty)^{-1/2}$[b]

[a]With L_∞ expressed in keV μm^{-1}
[b]For example, for $L_\infty = 60$ keV μm^{-1}, $Q = (0.32 \times 60) - 2.2$, or 17. All calculations of Q using the data in Table 4.2 should be rounded to the nearest whole number.

complications in personnel dosimetry. The Council also believes there is a reduced effectiveness of heavy ions with LET greater than 100 keV μm^{-1} as reflected in Table 4.2.

For its basic recommendations, however, the Council endorses the ICRP position that the detail and precision inherent in using a formal quality factor-LET relationship to modify absorbed dose to reflect the higher probability of detriment resulting from exposure to radiation components with high-LET is not justified because of the wide range of values in the radiobiological information as given in Table 4.1. In addition, the quality factor-LET relationship is based on the distribution of LET in a small volume of tissue (point). The Council now focuses its recommendations on the concept of the average dose in a specific tissue or organ. This results from the basic assumption of the linear hypotheses under which variations of dose within a tissue of uniform sensitivity to cancer induction is unimportant. It is for these reasons that the Council now recommends the use of w_R values.

The values for w_R for each specified radiation type and energy were chosen on the basis of a review of measured values of the RBE of the radiations for a variety of relevant biological effects at low absorbed doses, including those on human material when available.

The Council notes that derivations of "effective" $\overline{Q}$ values from calculations for a variety of radiations using the Q to L relationship given in Table 4.2 give values not very different from its selected values of w_R (see ICRP, 1991a, Annex A, Figures A-2 and A-3). The recommended values are given in Table 4.3.

The Council also endorses the ICRP approach to the calculations required for radiation types and energies which are not included in Table 4.3. For these cases, an approximation of w_R can be obtained by calculation of $\overline{Q}$ at 10 mm depth in the International Commission on Radiation Units and Measurements' sphere as given in Equation 4.3 below.

$$w_R \approx \overline{Q} = \frac{1}{D} \int_0^\infty Q(L) D(L) \, dL \qquad (4.3)$$

where $D(L)dL$ is the absorbed dose at 10 mm depth between linear energy transfer L and $L + dL$; and $Q(L)$ is the quality factor of L at 10 mm.

TABLE 4.3 — *Radiation weighting factor,* w_R.[a]
(Adapted from ICRP, 1991a.)

Type and energy range	w_R
X and γ rays, electrons, positrons and muons[b]	1
Neutrons, energy < 10 keV	5
10 keV to 100 keV	10
> 100 keV to 2 MeV	20
> 2 MeV to 20 MeV	10
> 20 MeV	5
Protons[c], other than recoil protons and energy > 2 MeV	2[d]
Alpha particles, fission fragments, nonrelativistic heavy nuclei	20

[a]All values relate to the radiation incident on the body or, for internal sources, emitted from the source.

[b]Excluding Auger electrons emitted from nuclei bound to DNA since averaging the dose in this case is unrealistic. The techniques of microdosimetry are more appropriate in this case.

[c]In circumstances where the human body is irradiated directly by >100 MeV protons, the RBE is likely to be similar to that of low-LET radiations and, therefore, a w_R of about unity would be appropriate for that case.

[d]The w_R value for high energy protons recommended here is lower than that recommended in ICRP (1991a).

5. Effective Dose

The effective dose (E) has associated with it the same probability of the occurrence of cancer and genetic effects whether received by the whole body *via* uniform irradiation or by partial body or individual organ irradiation. While an assumption of uniformity may be a sufficient approximation in many external irradiation cases, in others more precise evaluations of individual tissue doses will be necessary. With external irradiation, differences may arise with depth in the body and with orientation of the body in the generally nonuniform radiation field. *When irradiation is from radionuclides deposited in various tissues and organs, nonuniform or partial body exposures usually occur.* Tissues also vary in their sensitivity to radiation. The effective dose (E) is a concept similar to the effective dose equivalent (H_E) used by ICRP (1977) and NCRP (1987).[4] However, they are conceptually different (also see Section 4.1 regarding the difference between equivalent dose and dose equivalent). The effective dose (E) is intended to provide a means for handling nonuniform irradiation situations, as did the earlier effective dose equivalent.

The effective dose (E) is the sum of the weighted equivalent doses for all irradiated tissues or organs. The tissue weighting factor (w_T) takes into account the relative detriment to each organ and tissue including the different mortality and morbidity risks from cancer, the risk of severe hereditary effects for all generations, and the length of life lost due to these effects. The risks for all stochastic effects will be the same whether the whole body is irradiated uniformly or nonuniformly if

$$E = \sum_T w_T H_T , \qquad (5.1)$$

---

[4]The symbol E is used for the effective dose in accordance with ICRP (1991a).

where w_T is the tissue weighting factor representing the proportionate detriment (stochastic) of tissue T when the whole body is irradiated uniformly, and H_T is the equivalent dose received by tissue T.

Values of w_T recommended by the ICRP (1991a) and by the NCRP (1993a) are adopted for the purposes of the recommendations in this Report. These values are given in Table 5.1. The organ risks from which they were derived are given later in Table 7.1. The ICRP (1991a) showed that for the evaluation of the relative contribution of cancer in various organs to the total cancer risk, it is evident that the model used for the transfer of risks from one population to another, as well as the special characteristics of some national populations, can be more important than variables such as sex and age. However, in the interest of uniformity, while recognizing the uncertainties involved, the NCRP uses the same estimates that the ICRP (1991a) has used for fatal cancer risk and aggregated detriment, both in total and for individual organs. [For discussion, see NCRP Report No. 115 (NCRP, 1993a).]

The probability of fatal cancer and severe genetic effects and the total detriment weighted for length of life lost and with a nonfatal cancer component of detriment included are listed by organ in Table 7.2. The values of w_T are rounded and simplified values developed for a reference population of equal numbers of both sexes and a wide range of ages. Therefore, they should not be used to obtain specific estimates of potential health effects for a given individual.

Two axioms inherent in the selection and application of w_R and w_T values are:

(1) w_R is independent of the tissue or organ and

(2) w_T is independent of the radiation type or energy, *i.e.*,

$$E = \sum_R w_T D_{T,R} = \sum_T w_T \sum_R w_R D_{T,R} . \qquad (5.2)$$

TABLE 5.1 — *Tissue weighting factor (w_T) for different tissues and organs.*[a]
(Adapted from ICRP, 1991a and NCRP, 1993a.)

0.01	0.05	0.12	0.20
Bone surface	Bladder	Bone marrow	Gonads
Skin	Breast	Colon	
	Liver	Lung	
	Esophagus	Stomach	
	Thyroid		
	Remainder[b,c]		

[a]The values have been developed for a reference population of equal numbers of both sexes and a wide range of ages. In the definition of effective dose, they apply to workers, to the whole population and to either sex. These w_T values are based on rounded values of the organ's contribution to the total detriment.

[b]For purposes of calculation, the remainder is composed of the following additional tissues and organs: adrenals, brain, small intestine, large intestine, kidney, muscle, pancreas, spleen, thymus and uterus. The list includes organs which are likely to be selectively irradiated. Some organs in the list are known to be susceptible to cancer induction. If other tissues and organs subsequently become identified as having a significant risk of induced cancer, they will then be included either with a specific w_T or in this additional list constituting the remainder. The remainder may also include other tissues or organs selectively irradiated.

[c]In those exceptional cases in which one of the remainder tissues or organs receives an equivalent dose in excess of the highest dose in any of the 12 organs for which a weighting factor is specified, a weighting factor of 0.025 should be applied to that tissue or organ and a weighting factor of 0.025 to the average dose in the other remainder tissues or organs [see ICRP (1991a)].

6. Committed Equivalent Dose, Committed Effective Dose, Annual Reference Levels of Intake and Derived Reference Air Concentrations

6.1 Committed Equivalent Dose, Committed Effective Dose

Radiation doses received from radionuclides deposited in organs and tissues will be distributed temporally depending upon the effective half-life of the radionuclide. To take account of this continuing irradiation of organs and tissues that occurs after the intake of radionuclides, the NCRP continues the use of the committed dose concept. The committed equivalent dose, $H_T(\tau)$, is the time integral of the equivalent dose-rate in a specific tissue (T) following intake of a radionuclide into the body. For a single intake of radionuclide at time t_0, $H_T(\tau)$ is given by Equation 6.1, where $\dot{H}_T$ is the relevant equivalent dose-rate in an organ or tissue T at time t and τ is the period of integration. Unless specified otherwise, an integration time of 50 y after the intake is recommended for the occupational case and 70 y for members of the public.

$$H_T(\tau) = \int_{t_0}^{t_0+\tau} \dot{H}_T \, dt \qquad (6.1)$$

The committed effective dose $E(\tau)$, for each internally deposited radionuclide is calculated by summing the products of the committed equivalent doses and the appropriate w_T values for all tissues irradiated. The general equation is:

$$E(\tau) = \sum_T w_T H_T(\tau). \qquad (6.2)$$

The specific equation, with $\tau = 50$ y, is:

$$E(50) = \sum_T w_T H_T(50). \qquad (6.3)$$

For radionuclides with approximate effective half-lives ranging up to about three months, the committed quantities are approximately equal to the annual quantities for the year of intake. For radionuclides with an effective half-life, exceeding three months, the committed equivalent dose and the committed effective dose are greater than the equivalent or effective dose received in the year of intake because they reflect the dose that will be delivered in the future as well as that delivered during the year of intake. For radionuclides with a long effective half-life in comparison with remaining years of life of the individual exposed, neither a full expression of the risk nor the total dose will be manifested. For this reason, the committed equivalent dose and the committed effective dose from the life-long intake of radionuclides of very long effective half-life will overestimate by a factor of approximately two, or more (NCRP, 1987), the lifetime equivalent dose or effective dose. These quantities, therefore, are not particularly useful for estimating health effects or assessing probability of causation. However, the committed equivalent dose and the committed effective dose are appropriate for all routine radiation-protection purposes and should be used, for example, for assessing compliance with the annual effective dose limits and for planning and design. The annual effective dose limit referred to here is the sum of the external effective dose and the committed effective dose from internal emitters.

6.2 Annual Reference Levels of Intake: Occupational

The Annual Limits on Intake (ALI) given by ICRP (1991b) are based on limiting the committed effective dose from an intake in a single year to 20 mSv. The NCRP recommends the use of the ICRP values as reference values (see Section 13) rather than limits since intakes up to 2.5 times the ALI would be in compliance with the effective dose limit of 50 mSv given in Section 8. However, since the NCRP lifetime limit of age × 10 mSv (see Section 8) and an annual exposure of 20 mSv protects individual tissues against the likelihood of deterministic effects, Annual Reference Levels of Intake (ARLI) based on 20 mSv are adopted.

If the behavior of any specific material is expected to vary significantly from that of the dosimetric model employed, then adjustments should be made in the application of the model when specific data are available.

A useful alternative to the use of the ARLI is to obtain the committed effective dose per unit intake, *i.e.*, 20 mSv divided by the ARLI in becquerels. This information, when combined with an estimate of the intakes by a given individual, will allow for a direct summation with the external effective dose to assess compliance with the annual effective dose limits.

6.3 Derived Reference Air Concentrations

The Derived Reference Air Concentration (DRAC) is that concentration of a radionuclide which, if breathed by Reference Man, inspiring 0.02 m^3 per min for a working year, would result in an intake of one ARLI. Thus, the DRAC is determined by dividing the ARLI by 40 h per week, 50 weeks per y, 60 min per h and 0.02 m^3 per min.

$$DRAC = \frac{ARLI}{40 \text{ h week}^{-1} \times 50 \text{ week y}^{-1} \times 60 \text{ min h}^{-1} \times 0.02 \text{ m}^3 \text{ min}^{-1}} \qquad (6.4)$$

The purpose of the DRAC is to provide a method for controlling exposures in the workplace to the ARLI. Since the values for DRAC apply to individual radionuclides, they should be reduced appropriately for each radionuclide when two or more radionuclides are involved.

The DRAC calculated for workers cannot, of course, be used directly to control exposures of members of the public. Differences in factors such as applicable equivalent dose limits, duration of exposure, breathing rate, body size, metabolism and transfer factors would invalidate such use (ICRP, 1984). Further, exposures *via* other environmental pathways would have to be considered, food and water, for example. On the other hand, derived concentrations of radionuclides in water, for example, could be calculated when needed, in a manner similar to that employed for the DRAC, allowing for differences in dose limits and other variables such as those given above.

7. Risk Estimates for Radiation Protection

In Section 7 of NCRP Report No. 91 (NCRP, 1987), it was pointed out that although the nominal risk estimates of 1977 (ICRP, 1977; UNSCEAR, 1977) were still in use for radiation protection, it was already evident that the 1977 risk estimates would be revised to higher values. These values could not be stated in 1987 but the potentially higher values clearly influenced the tone and the guidelines given in NCRP Report No. 91 (NCRP, 1987).

It is now possible to be somewhat more definitive about risk estimates for cancer and for genetic effects even though many uncertainties still remain, including the magnitude of the neutron component used in the DS86 analysis. The data obtained from the study of Japanese survivors of the atomic bombs have been evaluated in separate reviews including those by investigators at the Radiation Effects Research Foundation in Hiroshima, e.g., Preston and Pierce (1988), Shimizu et al. (1987; 1990); by UNSCEAR (1988) and by the BEIR V Committee (NAS/NRC, 1990). The UNSCEAR (1988) and BEIR V Reports separately considered all other sources of human epidemiological information as well, and concluded that the Japanese survivors provided by far the most complete data source for external low-LET radiation and that risk estimates derived from them were broadly supported by the results of other studies. The reviews produced estimates of lifetime cancer risk for the general population after high dose and high-dose rate exposure ranging from about 9 to about 12×10^{-2} Sv^{-1} based on multiplicative or modified multiplicative projection models. The UNSCEAR and BEIR committees were not specific as to how to convert the risk to low dose (or low-dose rate) low-LET radiation exposure but suggested dividing the numerical values by two to ten (UNSCEAR, 1988) or two or more (NAS/NRC, 1990). Both committees estimated the genetic risk but did not provide a risk estimate for multifactorial diseases.

The ICRP, in their recent assessment (ICRP, 1991a), concluded that it would be appropriate to use a nominal value of 10×10^{-2} Sv^{-1} effective dose for the lifetime risk of fatal cancer for a population of all ages and 8×10^{-2} Sv^{-1} effective dose for a working population, for high dose, high-dose rate exposure. The ICRP (1991a) assessment was based on UNSCEAR (1988) and NAS/NRC (1990). After considering various experimental and human information, the ICRP also chose a Dose and Dose-Rate Effectiveness Factor (DDREF) of two, to convert risk estimates after high dose and high-dose rate exposure to those to be expected after low dose or low-dose rate exposure. Thus, the nominal values of lifetime cancer risk for low dose or low-dose rate exposure were stated to be 5×10^{-2} Sv^{-1} for a population of all ages and 4×10^{-2} Sv^{-1} for a working population (ICRP, 1991a).

The Council's assessment of risk for radiation-protection purposes is set out in another report (NCRP, 1993a). In that report, it is determined that for a United States population, the nominal values of lifetime risk of fatal cancer for a working population can be taken as 8×10^{-2} Sv^{-1} and 10×10^{-2} Sv^{-1} for a population of all ages for high dose, high-dose rate exposure, the same values as those used by ICRP (1991a). The choice of DDREF is somewhat arbitrary and the NCRP considered that it could reasonably range between two and three. Thus, nominal values of lifetime risk for low dose or low-dose rate exposure could range between 3.3 to 5×10^{-2} Sv^{-1} for a population of all ages and 2.7 to 4×10^{-2} Sv^{-1} for a working population. The differences between the values in these ranges are not significant. Therefore, the NCRP (1993a) recommends the use of 4×10^{-2} Sv^{-1} for workers and 5×10^{-2} Sv^{-1} for the general population for the lifetime risk of fatal cancer, the same values as those recommended by the ICRP, thereby endorsing a dose-rate effectiveness factor of two. These values are used in this Report.

Although recent evidence from studies of genetic damage among atomic bomb survivors suggests that humans are less sensitive to genetic effects than previously thought (NAS/NRC, 1991), for the derivation of dose limits and w_T values, a risk value for severe hereditary effects of 1×10^{-2} Sv^{-1} for all generations is recommended

(NCRP, 1993a). This includes the uncertain multifactorial component.[5] The assessment of the detriment resulting from severe hereditary effects set out in the NCRP report on risk estimates for radiation protection (NCRP, 1993a) includes derivation of lifetime values of 0.8×10^{-2} Sv^{-1} for workers and 1.3×10^{-2} Sv^{-1} for the whole population, after adjusting for length of life lost.

The ICRP (1991a) made estimates of total detriment, which included fatal cancer risks, nonfatal cancer risks and the risks of severe genetic effects modified by an adjustment according to the relative length of life lost. This detriment (equivalent fatal cancer risk) totaled 7.3×10^{-2} Sv^{-1} for a population of all ages and 5.6×10^{-2} Sv^{-1} for a working population.

In the assessment of total detriment set out in the NCRP risk estimate report (NCRP, 1993a), an estimate of lifetime risk of 5.6×10^{-2} Sv^{-1} for workers and 7.3×10^{-2} Sv^{-1} for the whole population was developed. Since these values are the same as those of ICRP (1991a), this Report recommends that the ICRP values of total detriment be used (see Table 7.1). The w_T values given in Table 5.1 are based on rounded values of their relative contribution to the total detriment.

Recent analyses of data from the Japanese atomic bomb survivors have been published based on incidence, rather than mortality data (Thompson *et al.*, 1992). These analyses provide more detail on cancer risks than possible with mortality data and may be considered in future evaluations.

The probability of fatal cancer and severe genetic effects and the total detriment weighted for length of life lost are listed by organ in Table 7.2 [see ICRP (1991a) and the NCRP report on risk estimates (NCRP, 1993a)]. The total detriment includes a nonfatal cancer component.

---

[5]Estimates of the multifactorial component must be considered highly uncertain at this time. The ICRP based its calculations on the incidence of multifactorial diseases as given in UNSCEAR (1988), the mutation component as given in UNSCEAR (1982), and a reduction factor for severity. If the same procedure were adopted, but the incidence of the multifactorial diseases given in the BEIR V Report (NAS/NRC, 1990) and the mutation component given in the BEIR III Report (NAS/NRC, 1980) were used, a substantially higher estimate of the multifactorial component would be obtained.

TABLE 7.1 — *Nominal probability coefficients for stochastic effects.*
(Adapted from ICRP, 1991a and NCRP, 1993a.)

Exposed Population	Detriment			
	Fatal Cancer[a] $(10^{-2} \ Sv^{-1})^b$	Nonfatal Cancer $(10^{-2} \ Sv^{-1})^b$	Severe genetic Effects $(10^{-2} \ Sv^{-1})^b$	Total Detriment $(10^{-2} \ Sv^{-1})^b$
Adult workers	4.0	0.8	0.8	5.6
Whole population	5.0	1.0	1.3	7.3

[a]For fatal cancer, the detriment coefficient and the probability coefficient are equal since the total detriment is that resulting from the fatal cancer only.
[b]Rounded values.

Table 7.2 provides estimates of total lifetime detriment and probability of fatal cancer for workers and whole populations based on available risk information. Although appearing to be highly quantitative and indeed probably better defined than at any other point in time, these estimates of risk must be recognized as subject to many uncertainties. These include uncertainties:

(1) of an epidemiological nature (statistical, underreporting of cancer, etc.),

(2) in dosimetry (random error, bias and other errors),

(3) in RBEs,

(4) in the projection of risks from the period of observation to total lifetime,

(5) in the transfer of risks between populations with different underlying cancer incidence rates and

(6) associated with extrapolation of risk data from high-dose rate exposure to low-dose rate exposure.

As an example of the potential concerns about uncertainties, one might turn to the matter of the transfer of risks between United States and Japanese populations, *e.g.*, the likelihood that nearly fifty percent of excess fatal cancers in a United States population following a 1 Sv low-dose rate exposure would be due to cancers of the gastrointestinal sites (esophagus, stomach and colon). While this might appear questionable, such estimates exist in the literature and have been accepted by authoritative bodies in spite of some reservations. Issues such as these, among others that could be raised over other potential uncertainties, serve to reinforce the Council's concern with regard to the degree of uncertainty in the available risk information, and emphasize the need for radiation protection bodies to exercise appropriate judgment in providing guidance on limits of exposure. It should be clear that the guidance provided in this Report is based on the Council's judgment regarding:

(1) the available data,
(2) other radiation protection considerations and
(3) our experience to date.

TABLE 7.2 — *Relative contribution of individual tissues and organs to the probability of fatal cancer and the total detriment.*[a]
(Adapted from ICRP, 1991a and NCRP, 1993a.)

	Probability of Fatal cancer (10^{-2} Sv^{-1})		Total detriment (10^{-2} Sv^{-1})	
	Whole Populations	Workers	Whole Populations	Workers
Bladder	0.30	0.24	0.29	0.23
Bone marrow	0.50	0.40	1.04	0.83
Bone surface	0.05	0.04	0.07	0.06
Breast	0.20	0.16	0.36	0.29
Esophagus	0.30	0.24	0.24	0.19
Colon	0.85	0.68	1.03	0.82
Liver	0.15	0.12	0.16	0.13
Lung	0.85	0.68	0.80	0.64
Ovary	0.10	0.08	0.15	0.12
Skin	0.02	0.02	0.04	0.03
Stomach	1.10	0.88	1.00	0.80
Thyroid	0.08	0.06	0.15	0.12
Remainder	0.50	0.40	0.59	0.47
Total	5.00	4.00	5.92	4.73
	Probability of Severe genetic effects			
Gonads	1.00	0.60	1.33	0.80
Grand Total			7.25	5.53

[a]The significant figures given in this Table are not given to imply a given degree of accuracy. Rather, they are carried forward to establish the tissue weighting factor (w_T).

8. Occupational Dose Limits

For the purpose of deriving the effective dose limit for occupational exposure, a single uniform whole body equivalent dose of 0.1 Sv is assumed to result in an average nominal lifetime excess risk of 4×10^{-3} for fatal cancer, 0.8×10^{-3} for severe genetic effects plus a nonfatal cancer risk of 0.8×10^{-3} for a total detriment of 5.6×10^{-3} (see Section 7).

In addition, if a death attributable to cancer occurs, it will result in a mean loss of approximately 15 y of life [see ICRP (1991a) and NCRP (1993a)]. This loss reflects the fact that more radiation-induced cancers, regardless of age at exposure, occur late in life. In contrast, accidental deaths at work may occur at any time in the working lifetime, resulting, on the average, in a greater number of years of life lost [see ICRP (1985) and NSC (1992)].

The average fatal accident rate in all industry is heavily influenced by the risk to workers in safe industry due to the higher proportion of workers in safer industries. That rate is approximately 1×10^{-4} y^{-1} (Table 3.1). The range of annual risk of accidental death in industry is about 0.2×10^{-4} to 5×10^{-4} and the mean age of death of those who suffer an accidental death in industry is approximately 40 y (ICRP, 1985). The data for 1980 (NCRP, 1989a) indicate that the average annual dose equivalent of monitored workers with measurable exposure was approximately 2.1 mSv which would suggest that the total detriment incurred by monitored workers (2.1×10^{-3} Sv y^{-1} × 5.6×10^{-2} detriment Sv^{-1}) is about 1×10^{-4} y^{-1}, which is consistent with the average risk of accidental death for all industries.

For those few individuals who might receive doses close to the limit over their working life, the Council believes that their total lifetime attributable detriment incurred each year should be no greater than the annual risk of accidental death for a worker at the top end of the safe worker range (between 10^{-4} and 10^{-3}).

In its 1987 recommendations (NCRP, 1987), the Council introduced the concept of a limitation of lifetime exposure based on age in the form of the following guidance, "*the community of radiation users is*

33

encouraged to control their operations in the workplace in such a manner as to ensure, in effect, that the numerical value of the individual worker's lifetime effective dose equivalent in tens of mSv (rem) does not exceed the value of his or her age in years." Now that risk estimates predicted in that report have been reflected in the UNSCEAR, NAS/NRC, ICRP and NCRP reviews, the Council believes that the guidance for lifetime exposure should be raised from guidance to the level of a basic recommendation.

The Council, therefore, recommends that the numerical value of the individual worker's lifetime effective dose in tens of mSv be limited to the value of his or her age in years (not including medical and natural background exposure). Exposures to individuals under age 18 shall be limited under the guidance given in Section 18. Clearly, this recommendation is not intended to suggest that it is acceptable that younger workers be allowed higher annual exposures than older workers simply by virtue of their age.

In order to control exposure more tightly in the early years of an individual's career and to provide flexibility in later years for those current operations or practices that may result in annual exposure to individuals in excess of 10 mSv, **the Council recommends that the annual occupational effective dose be limited to 50 mSv (not including medical and natural background exposure).** Under these two criteria (age × 10 mSv and 50 mSv per y), and using the worst case scenario, the lifetime fatal cancer risk would be approximately 3×10^{-2} (see Appendix A). The worst case scenario for accidental death in safe industry is 5×10^{-4} y^{-1} × 50 y which results in a lifetime fatal accident risk of 2.5×10^{-2}.

Alternatively, if the flexibility inherent in the above recommendations is not required for specific groups of workers, the implementation of an annual limit of 10 mSv is recommended.

The ICRP, in its Publication 60 (ICRP, 1991a), recommended a limit of 100 mSv in 5 y and no more than 50 mSv in 1 y. The overall objective of both the NCRP and ICRP dose-limit recommendations is to control the lifetime risk to the maximally exposed individuals. This is done by limiting lifetime irradiation to approximately 1 Sv in the case of ICRP and approximately 0.7 Sv in the case of NCRP (see Appendix A for a comparison of the risks associated with these recommendations). The NCRP recommendation on exposure limits provides somewhat greater flexibility in the control of worker exposures and requires the maintenance of individual cumulative exposure records. The ICRP system provides somewhat less flexibility and only requires the maintenance of exposure records over five year periods. The Council believes that cumulative lifetime exposure records are desirable for a number of reasons, *e.g.*, estimation of lifetime risks, epidemiological studies and long-term evaluation of exposures occurring in a given workplace [see NCRP Report No. 114, *Maintaining Radiation Protection Records* (NCRP, 1992)].

Since it is clear from the above discussion that the dose limit defines the edge of unacceptability, the upper bound nature of the dose limit is reemphasized. The advice given in NCRP Report No. 91 (NCRP, 1987) is still valid, *i.e.*, "It is only when the cost of further dose reduction is truly unreasonable that the limit should be approached."

The 50 mSv annual limit should be utilized only to provide the flexibility required for existing facilities and practices. **The NCRP recommends that all new facilities and the introduction of all new practices should be designed to limit annual exposures to individuals to a fraction of the 10 mSv per y limit implied by the cumulative dose limit.** Close attention to reducing exposures to ALARA is an integral part of ensuring that occupationally exposed workers enjoy the same level of protection as those in safe industry.

The annual effective dose limits apply to the sum of the effective dose from external irradiation and the committed effective dose from internal exposures.

9. Dose Limits for Deterministic Effects: Occupational

The deterministic effects of concern are those effects severe enough to be clinically significant (see Section 2).

As indicated in Section 6, the new ARLI values (ICRP, 1991b) are based on a 50-y committed effective dose limit of 20 mSv, which results in only a few radionuclides delivering lifetime doses that could approach deterministic levels. In addition, there is a reduced likelihood of deterministic effects from protracted low-LET irradiation. Use of the equivalent dose in calculating ARLI values to tissues or organs from alpha-emitting radionuclides also over-estimates deterministic effects since a w_R of 20 for alpha particles, based on the risk of stochastic effects, considerably exceeds the RBE of alpha particles for deterministic effects, which is certainly less than ten.

The Council, therefore, believes recommendations are required only for the crystalline lens of the eye, the skin, the hands and the feet.[6] The following annual equivalent dose limits are recommended for the occupational case: 150 mSv for the crystalline lens of the eye, and 500 mSv for localized areas of the skin, the hands and feet. These limits apply whether an individual tissue or organ is exposed selectively or together with other tissues or organs. Such discrete limits are required for the hands, for example, since limitation based on stochastic risks alone (effective dose) would permit several sieverts per year, which could eventually result in deterministic effects.

---

[6]The effective dose limit provides adequate protection for the skin against stochastic effects. An additional limit is needed for the skin in order to prevent deterministic effects.

10. Protection of the Embryo-Fetus

In rats and mice, irradiation of the embryo-fetus to substantial doses has been shown to produce a wide spectrum of developmental anomalies (UNSCEAR, 1986). The only excess incidence of anomalies that has been reported in detail in the atomic bomb survivors involves the central nervous system, small head size, severe mental retardation and deficits in intelligence (Otake and Schull, 1984). The absence of a wide spectrum of developmental anomalies in the Japanese survivors may reflect the very low LD_{50} for the human embryo, the very short period of sensitivity for the induction of central nervous system effects and/or that the doses may not have been high enough to induce other developmental anomalies.

Among atomic bomb survivors exposed *in utero*, a dose-dependent increase in the incidence of severe mental retardation occurred in the gestational age group of 8 to 15 weeks after conception and, to a lesser extent, in the gestational age group of 16 to 25 weeks after conception. Subjects exposed to radiation at less than 8 weeks or after 26 weeks of gestational age were not observed to have an excess of mental retardation. The data are consistent with a linear relationship between the incidence of mental retardation and dose with a slope of 0.4 Gy^{-1} and a threshold of 0.1 to 0.2 Gy, if all cases of severe mental retardation are included. If two children with Down's Syndrome are excluded (they clearly had a nonradiation etiology since their genetic condition existed prior to exposure), the threshold is about 0.4 Gy. These data strongly suggest a threshold for severe mental retardation even in the most sensitive stages of gestation. The relative risk for exposure during the 16 to 25 week period is at least four times less than that for exposure at 8 to 15 weeks after conception, with an even clearer indication of a threshold.

For exposures during the most sensitive period, 8 to 15 weeks, more recent data from Japan also indicate a reduction in intelligence scores of 21 to 29 points at 1 Gy, although the data show great variability (Otake *et al.*, 1988; Schull *et al.*, 1988).

Small head size was observed over a wider range of doses and in some cases over a wider range of gestational age than for severe mental retardation, although the majority of the cases cluster around the 8 to 15 week period. There is not a one-to-one correlation between small head size and severe mental retardation, since one may occur in the absence of the other, with small head size being more common than mental retardation.

Epidemiological data suggest an association between diagnostic x rays received *in utero* and an excess incidence of childhood cancer, which implies that susceptibility to radiation carcinogenesis from exposure during prenatal life may be higher than in the adult. However, there is also evidence against this conclusion. An excess in childhood cancer was not observed in the atomic bomb survivors that were exposed *in utero*. In a follow-up through 1984, cancer incidence data suggested that excess cancer in adulthood might result from irradiation *in utero*, although there was no evidence of excess cancer risk for more recent years, 1985-89 (Yosimoto *et al.*, 1992). The present data are consistent with a lifetime cancer risk resulting from exposure during gestation which is two to three times that for the adult (*i.e.*, about 1×10^{-1} Sv^{-1}) (UNSCEAR, 1986).

The sensitivity of the embryo-fetus for both mental retardation and cancer should be considered in all situations involving irradiation of the embryo-fetus. Therefore, for occupational situations, the **NCRP recommends a monthly equivalent dose limit of 0.5 mSv to the embryo-fetus (excluding medical and natural background radiation) once the pregnancy is known.** This is based on the philosophy that a monthly limit will control exposure during potentially sensitive periods of gestation.

The recommendation reflects the need to limit the total lifetime risk of leukemia and other cancers in individuals exposed *in utero*. At doses below the limit, all deterministic effects including small head size and mental retardation are expected to be negligible.

The Council no longer believes that specific controls are required for occupationally exposed women who are not known to be pregnant. Under the highly unlikely maximum exposure scenario (50 mSv before pregnancy is known), the potential impact on IQ (intelligence quotient) would be expected to be negligible since the period of enhanced sensitivity is 8 to 15 weeks and beyond. In addition, the lifetime cancer risk would be expected to be less than 5×10^{-3}, *i.e.*, the lifetime cancer risk resulting from exposure during gestation times the annual occupational exposure limit (1×10^{-1} Sv$^{-1} \times 5 \times 10^{-2}$ Sv). While it is clearly desirable to avoid such a risk, it is small compared to the other risks to the fetus. It should be noted that even if pregnancy goes unrecognized beyond the period of enhanced sensitivity, the reduction in IQ would be no more than one or two IQ points. As it is extremely unlikely that any worker would receive more than 50 mSv, and since the Council does not wish to make a recommendation that is unnecessarily discriminatory, the Council withdraws its previous recommendation regarding the occupational exposure of women not known to be pregnant.

Internally deposited radionuclides pose special problems for protection of the embryo-fetus because some radionuclides remain in the body for long periods of time, and their transfer, and the doses delivered to fetal organs are not well known. Therefore, it is important to limit the intakes of radionuclides by pregnant women so that the equivalent dose to the embryo-fetus would not exceed the recommended limit. For the present, the NCRP adopts the ICRP recommendation that the intake of radionuclides, once pregnancy is known, be limited to about one-twentieth of the values of the ARLI for workers.[7]

---

[7]It is recognized that under some conditions of exposure this would allow for a different total dose to the embryo-fetus from internally deposited radionuclides than that recommended for external exposure.

The NCRP has a scientific committee addressing the risks to the embryo-fetus that intake of radionuclides by the mother pose and the Council will continue to examine other information relevant to the making of recommendations for protection of the embryo-fetus.

11. Exposure in Excess of the Dose Limits: Occupational

The annual dose limits are intended to control the maximum lifetime risk incurred by an individual in any year and normally exposures should be well below the limit. Nevertheless, slightly exceeding the annual effective dose limit in a given year has little biological significance for the individual since the lifetime risk can be readily offset by a history of exposure below the dose limits in past years or reduced exposure in future years. The importance of such occurrences is that they call attention to what may be an inadequate system of radiation control. This would suggest that the decision on permitting individuals who have exceeded annual dose limits to return to worker status depends more upon the improvement and corrections in the control of radiation exposure in the workplace than on in-depth analysis of the worker's health status.

Because there are individuals who, under the previous recommendations of the Council, were permitted to accumulate exposures in excess of the new age-related limit, **the Council recommends that individuals whose cumulative effective dose exceeds the age related limit should be restricted in their exposures to no more than 10 mSv per y until the age related lifetime limit is met.** If additional flexibility is warranted, particularly for older workers whose effective dose may approach or exceed 10 mSv $\times$ age, then, on a case by case basis, with dialog between the employee and employer, an exposure limit of up to an average of 100 mSv in 5 y and 50 mSv in any year may be considered.

Annual Reference Levels of Intake (ARLI) are intended to control the intake of radioactive materials. Occasional intakes slightly greater than 2.5 times the ARLI will have little effect on the long-term health of an individual. However, for those cases in which the intakes greatly exceed 2.5 times the ARLI, more detailed analysis is required.

Since the ARLI values are calculated for the ICRP Reference Man, any estimate of potential health effects on a given individual must be based on organ-dose estimates derived from information specific to the nature of the intake, such as the physical and chemical properties of the radionuclide, estimates of body content by *in vivo* counting or excretion analysis, and the age and other physiological parameters of the exposed individual. The occurrence of radionuclide intakes that exceed 2.5 times the ARLI, in general, indicate that improvements in facilities or operating procedures are needed.

12. Dose Limits for Unusual Occupational Situations

In the development of the recommendations pertaining to exposure in the workplace, the average risk in safe industries has been used as the basis. The Council believes that most occupations and industries involving ionizing radiation can adhere to this risk level without undue difficulty, and many can operate at a much lower risk level due to the relatively low occupational exposure required to accomplish the work. There may exist, however, unusual occupational situations in which the worker population cannot carry out required functions under the recommended annual limits and, in which, risks other than radiation may be much larger than normal. Such special circumstances may require special limits and a different basis of comparison. For example, the radiation risks in these circumstances could be compared with those from the less safe industries such as public utilities and transportation, construction, agriculture and mining. In these situations, the main concern also would be for the accumulated exposure of individuals over their working lifetime and thus a career limit would be of special importance. Space missions of long duration, especially those in deep space and possibly some mining circumstances, are examples of such situations. **Therefore, the Council recommends that consideration be given to establishing special dose limits for those selected occupational groups requiring higher exposures to accomplish needed activities.**

13. Reference Levels: Occupational

As stated earlier, annual dose limits are not generally suitable in and of themselves for design and control purposes because ALARA is also a part of the radiation-protection system. A reference level is a selected numerical value of effective dose that is used for design or control purposes. For example, if it can be demonstrated that operations at a given facility can be carried out without exceeding a certain monthly level below that corresponding to the annual limits, the management might well establish a periodic reference level based on this information. Such a reference level would serve to keep employees cognizant of the obligation to maintain their exposure ALARA and draw attention to the constant need for good radiation-protection practices. If an exposure exceeded this value, a predetermined course of action could be taken, such as investigation, retraining, maintenance or repair. (It may be appropriate to specify different levels for each of these follow-up actions.) The establishment of reference levels is most properly done at each facility because the data required will mainly be site-specific. However, for the local control of intake of radionuclides in occupational situations, the Council recommends the use of an ARLI based on limitation of effective dose to not more than 20 mSv (see Section 6.2).

14. Guidance for Emergency Occupational Exposure

Normally, only actions involving life saving justify acute exposures that are significantly in excess of the annual effective dose limit. The use of volunteers for exposures during emergency actions is desirable. Older workers with low lifetime accumulated effective doses should be chosen from among the volunteers, whenever possible. Exposures during emergency actions that do not involve life saving should, to the extent possible, be controlled to the occupational dose limits. Where this cannot be accomplished, it is recommended that a limit of 0.5 Sv effective dose and an equivalent dose of 5 Sv to the skin be applied, which is consistent with ICRP recommendations (ICRP, 1991a).

When, for life saving or equivalent purposes the equivalent dose may approach or exceed 0.5 Sv to a large portion of the body in a short time, the workers need to understand not only the potential for acute effects but they should also have an appreciation of the substantial increase in their lifetime risk of cancer. If internally deposited radionuclide exposures are also possible, these should be taken into account.

15. Nonoccupational Dose Limits: Exposure of Individual Members of the Public

The development of recommendations on limitation of radiation exposure to the public is more difficult than that for the occupational case because comparisons such as those used in that case cannot be as explicit. Nevertheless, there are pertinent considerations upon which to base reasonable judgments. First, among these, are the risks associated with a given dose and the uncertainties associated with the risk estimates, as discussed in Section 7. Second, the manifold mortality risks faced by members of the public vary greatly, but are commonly in the range of about 10^{-4} to 10^{-6} per y and sometimes higher (see, for example, ICRP, 1977; Pochin, 1981; Sinclair, 1985; Wilson and Crouch, 1982). Not only do these risks exist, but they seem, depending somewhat on their nature, to be tolerated. Third, everyone is exposed to natural background radiation, and that, annually, is commonly about 1 mSv (excluding radon) or an assumed annually incurred lifetime risk of mortality of about 10^{-4} to 10^{-5}. The annual total effective dose from natural sources (excluding radon) varies in the United States from about 0.65 mSv on the Atlantic Seaboard to 1.25 mSv in Denver, Colorado. The average annual effective dose due to radon is about 2 mSv and variations in it are much greater (NCRP, 1984a) than the average value of natural background from other sources.

On the basis of the above considerations and the increased potential period of exposure and wider range of sensitivities to be found in the general population, the limits recommended for the worker are reduced by a factor of ten for application to the public. Therefore, the following limits are recommended for exposures from man-made sources other than medical and natural background.[8] **For continuous (or frequent) exposure, it is recommended that the annual effective dose not exceed 1 mSv.** This recommendation is designed to limit the exposure of members of the public to reasonable levels of risk comparable with risks from other common sources, *i.e.*, about 10^{-4} to 10^{-6} annually. Furthermore, **a maximum annual effective dose limit of 5 mSv is recommended to provide for infrequent annual exposures** (NCRP, 1984a). An annual effective dose limit recommendation of 5 mSv is made because annual exposures in excess of the 1 mSv recommendation, usually to a small group of people, need not be regarded as especially hazardous, provided it does not occur often to the same groups and that the average exposure to individuals in these groups does not exceed an average annual effective dose of about 1 mSv.

This annual limit for continuous exposure implies, for the design of new facilities or the introduction of new practices, that the radiation protection goal in such cases should be that no member of the public would exceed the 1 mSv annual effective dose limit from all man-made sources not including exposures associated with his or her medical care.

Both the 1 mSv and the 5 mSv annual effective dose limits for members of the public will keep the annual equivalent dose to those organs and tissues that are considered in the effective dose system below levels of concern for deterministic effects. However, because some organs and tissues are not necessarily protected against deterministic effects in the calculation of effective dose, *i.e.*, hands and feet, skin and lens of eye (see Section 5), an annual equivalent dose limit of 50 mSv is recommended for the hands, feet and skin and 15 mSv is recommended for the lens of the eye.

Information on the effective dose equivalent per unit intake (Sv Bq^{-1}) as a function of age for a selected group of radionuclides is provided in ICRP Publication 56 (ICRP, 1989b). Using these values and the current w_T values and the annual effective dose limit given above, one can estimate the quantity of a radionuclide that would result in a committed effective dose equal to the dose limit as a function of age.

---

[8]Medical exposures refer to those exposures received by individual patients as a result of medical procedures performed on them and for which they anticipate a medical benefit.

These recommendations for members of the public apply to exposures from man-made sources and exclude medical and natural background exposures. Medical exposures are excluded from this limitation because they are assumed to result in personal benefit to the exposed individual. Guidelines exist for the exposure of the public to medical procedures (EPA, 1978; FDA, 1985a; 1985b). Natural background exposure is excluded also. Natural background radiation is ubiquitous and variations in natural background levels do not exceed a factor of two or three (except for radon). The first recommendation effectively limits the exposure of an individual from man-made sources to the same value as that from average natural background (excluding radon). Natural background exposure is, however, included in recommendations specifying remedial action levels (see Section 16).

Exposures controlled to the annual limit of 1 mSv effective dose for continuous and 5 mSv effective dose for infrequent exposures are subject to the application of ALARA techniques in the same way as for occupational exposures.

When exposures are from internal and external sources, the contributions should be summed in a manner similar to that given in Section 8 for occupational exposures, so that the total effective dose does not exceed the respective limits.

In the application of the Council's recommendations to sources irradiating members of the public, the overriding considerations are those of JUSTIFICATION and ALARA. Normally, application of these two principles will insure that individuals are adequately protected. However, the NCRP reaffirms its previous recommendations (NCRP, 1984b) that whenever the potential exists for exposure of an individual member of the public to exceed 25 percent of the annual effective dose limit as a result of irradiation attributable to a single site, the site operator should ensure that the annual exposure of the maximally exposed individual, from all man-made exposures (excepting that individual's medical exposure), does not exceed 1 mSv on a continuous basis. Alternatively, if such an assessment is not conducted, no single source or set of sources under one control should result in an individual being exposed to more than 0.25 mSv annually.

16. Remedial Action Levels for Naturally Occurring Radiation for Members of the Public

If the recommendations of the previous Section are observed, man-made radiation sources will not expose members of the public to annual effective doses greater than 1 mSv continuously, or 5 mSv infrequently. Exposures should always be less than the limits and, indeed, on the average, utilizing the principles of ALARA, they should be much less.

However, natural background is excluded from the above limits and there are circumstances in which natural background itself, or more especially, natural radiation sources enhanced locally by man's operations for selected purposes, can give rise (sometimes quite inadvertently) to annual exposures above the level of 1 mSv.

It then becomes necessary to consider at what exposure level remedial action, which may be possible only at substantial societal cost, should be undertaken. Remedial action levels involve a balance of risk and many other socio-economic factors. In general, the aim of setting a remedial action level is to reduce the greatest risks from a given type of radiation source. It is clear that once a remedial action level is established for given circumstances, action is recommended when a level above it is found. Actions to reduce exposure should not be limited by or to the remedial action level and, following the ALARA principle, levels substantially below the remedial action level may be obtainable and appropriate.

The NCRP has given special attention to the problems occasioned by exposure to indoor radon (NCRP, 1984a; 1984c; 1988; 1989b; 1991) and notes that this is potentially the most important public radiation-exposure problem that currently exists. As a result, a remedial action level at an annual exposure of 2 WLM or 7×10^{-3} Jh m^{-3} is recommended, a value ten times the average level found in United States' homes (NCRP, 1984a).[9] Elements of feasibility enter the considerations here since it is evident (NCRP, 1984a) that in a substantial number of homes the radon levels are estimated to exceed *the average* by amounts up to five or ten times or more. It is certainly desirable that such levels be reduced and the risks associated with them decreased. A remedial action level must, therefore, be chosen for which the greatest risks are avoided but the societal impacts are not excessive. The NCRP recognizes that an annual inhalation level for radon that corresponds to approximately 5 mSv effective dose would be about 1.75×10^{-3} Jh m^{-3} (see ICRP, 1981).[10] However, this is only two and one-half times the present estimated average annual indoor radon background exposure of 7.0×10^{-4} Jh m^{-3} and imposition of a remedial action level at this value could involve a very large number of homes and great societal cost. Therefore, the NCRP has proposed a remedial action level which is based on excess lifetime risk being no more than ten times the excess lifetime risk associated with the average annual background level found in homes that is 7.0×10^{-3} Jh m^{-3} y^{-1} (NCRP, 1984a). The annual risk of fatal lung cancer associated with an exposure of 7.0×10^{-3} Jh m^{-3} y^{-1} is assumed to be 4×10^{-4}, which is higher than the risks associated with the limits for other radiation sources (uncertainties associated with risk estimates are discussed in Section 7). However, the NCRP anticipates that, once taken, remedial actions, together with ALARA, will result in annual radon exposures in a given home considerably below 7.0×10^{-3} Jh m^{-3} y^{-1}.

It is also anticipated that, over time, and assuming that the problem of indoor radon is addressed by taking the worst situations first, radon levels in existing homes will be reduced and that ultimately a lower

[9]3.5×10^{-3} Jh m^{-3} is equal to 1 Working Level Month (WLM). 1 WLM is approximately equal to an annual exposure to an average of 4 pCi per liter of radon if the radon decay products are in 50 percent equilibrium with the radon. 1 WLM exposure would result from being exposed to 1 Working Level (WL) for a period of one working month *i.e.*, 170 hours. 1 WL is defined as that concentration of radon daughters which has a potential alpha energy release of 1.3×10^{5} MeV per liter (2×10^{-5} J m^{-3}) of air, see Appendix B, NCRP Report No. 97 (NCRP, 1988).

[10]This assumes that the values of dose equivalent and equivalent dose to the lungs from radon decay products are approximately the same.

remedial action level may be reasonable. Furthermore, the Council believes that for new homes, suitable construction constraints should be developed so that they will have radon levels below those of many present structures.

For the present, **it is recommended that remedial action for radon be undertaken when the total exposure to radon decay products for an individual exceeds an annual average of 7×10^{-3} Jh m^{-3}** (see Footnote 9).

In the case of other exposure from natural radiation sources, considerations similar to those applied to radon would appear to be reasonable. Since the average exposure to individuals in the United States from natural radiation sources, excluding radon, is approximately 1 mSv annually, **it is recommended that remedial action be undertaken when continuous exposures from natural sources, excluding radon, are expected to exceed five times the average, or 5 mSv annually.**

17. Negligible Individual Dose

The concept of a Negligible Individual Risk Level (NIRL) was introduced in 1987 (NCRP, 1987) and was defined as the level of average annual excess risk of fatal health effects attributable to radiation below which efforts to reduce radiation exposure to the individual is unwarranted. In deriving the recommended value of the NIRL, several criteria relevant to the low level of risk or triviality of risk were considered which, taken together, offer degrees of reasonableness and perspective that tend to minimize subjective aspects of judgment. Smallness of risk was considered in relation to:

(1) magnitude of dose,
(2) difficulty in detection and measurement of dose and health effects,
(3) natural risk for the same health effects,
(4) estimated risk for the mean and variance of natural background radiation exposure levels,
(5) risk to which people are accustomed and
(6) perception of, and behavioral response to, risk levels.

The limiting of radiation risk among those exposed in the workplace to levels of risk that are generally regarded as safe is a valuable objective approach. This kind of approach at the lower level of comparable risk contributes to the establishment of a value of the NIRL.

Based on the these criteria, the Council, in 1987, adopted an NIRL of 10^{-7} y^{-1} and noted that this corresponded to an annual effective dose equivalent of 0.01 mSv. The nominal risk of fatal health effects used in developing that corresponding effective dose equivalent of 0.01 mSv was 10^{-2} Sv^{-1} (actually about 1.25×10^{-2} Sv^{-1}) for fatal cancers and 0.4×10^{-2} Sv^{-1} for genetic effects for the first two generations for a total of 1.65×10^{-2} Sv^{-1}. As seen in Section 7 of this Report, the Council has now adopted a risk value for the general public of 5×10^{-2} Sv^{-1} for fatal cancers, 1.3×10^{-2} Sv^{-1} for serious genetic effects and 1.0×10^{-2} Sv^{-1} for the detriment associated with nonfatal cancers, for a total detriment of 7.3×10^{-2} Sv^{-1} (Table 7.1). This detriment includes factors additional to those used in NCRP Report No. 91 (NCRP, 1987). This might suggest that the corresponding annual effective dose equivalent of 0.01 mSv, developed by NCRP in 1987, should be reduced by about a factor of four or five.

However, a review of several of the considerations given above suggests that a reduction in the 1987 recommendation of 0.01 mSv may not be warranted. For example, "magnitude of the dose," "difficulty in detection and measurement of dose and health effects," would still apply to the 0.01 mSv, and "the estimated risk for the mean and variance of natural background radiation exposure levels" could have resulted in an increase by a factor of four or five or more in the initial selection of the NIRL. **The Council, therefore, recommends that an annual effective dose of 0.01 mSv be considered a Negligible Individual Dose (NID) per source or practice.** Clearly, this dose could also be utilized as the limiting value in screening methods given in NCRP Commentary No. 8 (NCRP, 1993b).

In its Report No. 91 (NCRP, 1987), the Council also recommended that assessments of increments of collective annual dose from any particular individual source or practice should exclude those individuals whose annual effective dose equivalent from such a source was 0.01 mSv or less. The Council withdraws this statement as a formal recommendation. Although the Council fully endorses the nonthreshold hypothesis for the purpose of radiation protection, it wishes to point out that making an assessment of collective dose when the individual doses are less than 0.01 mSv may not be cost effective. From another point of view, we can not exclude the possibility of a fatal cancer attributable to radiation in a very large population of people exposed to very low doses of radiation, but the same could be said for many other unregulated exposures; moreover, at very low levels of exposure, the societal impact could be considered to be negligible.

18. Individuals Exposed Under 18 Years of Age

For educational and training purposes, it may be necessary and desirable to accept occasional exposure of persons under the age of 18 y. It is recommended that exposures for these purposes be permitted only under conditions presenting high assurance of maintaining the resulting annual effective dose to less than 1 mSv and dose equivalent to the lens of the eye, to less than 15 mSv and to the hands, feet and skin to less than 50 mSv (excluding medical and natural background radiation exposure). This is considered to be a part of the annual limit of 5 mSv given in Section 15 for infrequent exposure for members of the public and not supplemental to it. Intentional exposure of trainees should be avoided.

It is recognized that a productive part of the training experience may be better conducted in an industrial or hospital situation, which might constitute part-time work experience, supervised in some manner by an educational institution. The NCRP recommends that such work experience be governed by the radiation-protection practices recommended for educational institutions in NCRP Report No. 32 (NCRP, 1966) with, however, the revision in limits recommended here.

19. Summary of Recommendations

A number of modifications of the Council's earlier recommendations on limits for exposure to ionizing radiation (NCRP, 1987) have been presented. These modifications and a summary of the current recommendations are presented here.

For occupational exposures:

(1) The NCRP recommends that the cumulative effective dose not exceed the age of the individual in years $\times$ 10 mSv (see Section 8).

(2) The NCRP continues the use of the annual limit of 50 mSv but only as a limit on the annual rate of effective dose (see Section 8).

(3) For the exposure of pregnant women under occupational conditions, it is recommended that there be a limitation on the rate of equivalent dose to the embryo-fetus of no more than 0.5 mSv in a month (see Section 10). With this recommendation, there is no need for a limit on the total equivalent dose received by the embryo-fetus.

(4) The NCRP continues to recommend explicitly that all dose limits apply to the sum of external and internal exposures; the external exposures being assessed through the effective dose and the internal exposures assessed on the basis of the committed effective dose (see Section 6).

(5) New facilities and the introduction of new practices should be designed to limit annual effective doses to workers to a fraction of the 10 mSv y^{-1} implied by the lifetime dose limit (see Section 8).

(6) In this Report the NCRP introduces Annual Reference Levels of Intake (ARLI) at the same effective dose level as recommended by the ICRP for Annual Limits on Intake (ALI) of 20 mSv for workers (see Section 6).

For public exposures:

(1) The NCRP recommends an annual limit of 1 mSv effective dose for continuous exposure and an annual limit of 5 mSv effective dose for infrequent exposures (not including medical or background exposures in either case) (see Section 15).

(2) The NCRP recommends remedial action levels for the public of 5 mSv annual average effective dose for exposure from natural sources excluding radon and an annual average of 7×10^{-3} Jh m^{-3} for total exposure to radon and its decay products (see Section 16).

Another important change is the introduction of the radiation weighting factors (w_R) which range from 1 for all photon energies up to 20 for 1 MeV neutrons and alpha particles. Also, in the interest of providing a uniform approach to radiation protection, the new definitions and concepts given in ICRP Publication 60 (ICRP, 1991a) have been adopted wherever practical.

In this Report, the NCRP defines an annual Negligible Individual Dose (NID) which establishes a boundary below which the dose can be dismissed from consideration and sets the NID at 0.01 mSv effective dose. The current recommendations on limits are summarized in Table 19.1. A comparison of these recommendations with those made by ICRP in Publication 60 (ICRP, 1991a) and the earlier recommendations of the NCRP (1987) is provided in Table 1.1.

TABLE 19.1 — *Summary of recommendations.*[a,b]

A. Occupational exposures[c]	
1. Effective dose limits	
a) Annual	50 mSv
b) Cumulative	10 mSv × age
2. Equivalent dose annual limits for tissues and organs	
a) Lens of eye	150 mSv
b) Skin, hands and feet	500 mSv
B. Guidance for emergency occupational exposure[c]	(see Section 14)
C. Public exposures (annual)	
1. Effective dose limit, continuous or frequent exposure[c]	1 mSv
2. Effective dose limit, infrequent exposure[c]	5 mSv
3. Equivalent dose limits for tissues and organs[c]	
a) Lens of eye	15 mSv
b) Skin, hands and feet	50 mSv
4. Remedial action for natural sources:	
a) Effective dose (excluding radon)	>5 mSv
b) Exposure to radon decay products	$>7 \times 10^{-3}$ Jh m^{-3}
D. Education and training exposures (annual)[c]	
1. Effective dose limit	1 mSv
2. Equivalent dose limit for tissues and organs	
a) Lens of eye	15 mSv
b) Skin, hands and feet	50 mSv
E. Embryo-fetus exposures[c] (monthly)	
1. Equivalent dose limit	0.5 mSv
F. Negligible individual dose (annual)[c]	0.01 mSv

[a]Excluding medical exposures.
[b]See Tables 4.2 and 5.1 for recommendations on w_R and w_T, respectively.
[c]Sum of external and internal exposures but excluding doses from natural sources.

APPENDIX A.

Comparison of the Fatal Cancer Risk Associated with Occupational Dose Limits Specified in ICRP Publication 60 and this Report

The ICRP, in its Publication 60 (ICRP, 1991a), recommends an occupational limit of 100 mSv effective dose in 5 y and no more than 50 mSv effective dose in any 1 y. The risks associated with these recommendations are calculated below using two scenarios, one considering uniform exposure and the other maximizing the exposure over the early years, specifically:

(A) uniform exposure at the rate of 20 mSv y^{-1} from age 18 through 64 (ICRP Scenario A, see Table A.1) and

(B) 50 mSv received at age 18 and age 19 and 20 mSv y^{-1} from age 23 through 64 (ICRP Scenario B, see Table A.1).

In this Report, the NCRP recommends that the individual worker's lifetime effective dose be limited to his or her age in years × 10 mSv effective dose and that the annual effective dose be limited to 50 mSv. The risks associated with these recommendations are calculated below using two scenarios, one considering uniform exposure and the other maximizing the exposure over the early years, specifically:

(A) uniform exposure at the rate of 13.6 mSv y^{-1} from age 18 through 64 (NCRP Scenario A, see Table A.1) and

(B) 50 mSv received each year at age 18 through 21, 20 mSv at age 22 and 10 mSv y^{-1} from age 23 through 64 (NCRP Scenario B, see Table A.1).

These estimates of lifetime excess risk were calculated from intermediate results obtained by Land and Sinclair (1991). Risks were averaged over the sexes and between the multiplicative and NIH models, and a DDREF of two was applied.

TABLE A.1 — *Cumulative risks for NCRP and ICRP exposures at the limits.*

Scenario			Cumulative Exposure	Cumulative risk × 10^{-2}		
				M	F	Ave
NCRP	(A)	Uniform 13.6 mSv y^{-1} Age 18-64	640 mSv	2.1	2.8	2.5
	(B)	"Worst Case" 50 mSv y^{-1} Age 18-21 20 mSv y^{-1} Age 22 10 mSv y^{-1} Age 23-64	640 mSv	2.6	3.6	3.1
ICRP	(A)	Uniform 20 mSv y^{-1} Age 18-64	940 mSv	3.1	4.1	3.7
	(B)	"Worst Case" 50 mSv y^{-1} Age 18-19 20 mSv y^{-1} Age 23-64	940 mSv	3.3	4.4	3.9

Glossary

absorbed dose: The quotient of $d\bar{\epsilon}$ by dm where $d\bar{\epsilon}$ is the mean energy imparted by ionizing radiation to the matter in a volume element and dm is the mass of the matter in that volume element, $i.e.$, the absorbed dose, $D = d\bar{\epsilon}/dm$. The unit of absorbed dose is the gray (Gy).

Annual Reference Levels of Intake (ARLI): The activity of a radionuclide that, taken into the body during a year, would provide a committed effective dose to a person, represented by Reference Man, equal to 20 mSv. The ARLI is expressed in becquerels (Bq).

becquerel (Bq): The special name for the unit of activity. $1 \text{ Bq} = 1 \text{ s}^{-1}$.

committed effective dose $E(\tau)$: The committed equivalent doses to individual tissues or organs resulting from an intake multiplied by the appropriate tissue weighting factor (w_T) and then summed. $E(\tau) = \Sigma \, w_T H_T(\tau)$ where $H_T(\tau)$ is the committed equivalent dose in tissue T, w_T is the weighting factor for tissue T and τ is the integration period in years.

committed equivalent dose $H_T(\tau)$: The equivalent dose in a particular organ or tissue accumulated in a specified period τ, after intake of a radionuclide. It is defined by:

$$H_T(\tau) = \int_{t_0}^{t_0+\tau} \dot{H}_T(\tau)$$

where $\dot{H}_T(\tau)$ is the equivalent dose rate in an organ or tissue T at time t and τ is given in years, $i.e.$, $\tau = 50$ y is applicable to workers and $\tau = 70$ y is applicable to members of the public.

Derived Reference Air Concentrations (DRAC): The ARLI of a radionuclide divided by the volume of air inhaled by Reference Man in a working year ($i.e.$, $2.4 \times 10^3 \text{ m}^3$). The unit of DRAC is Bq m^{-3}.

deterministic effects: Effects for which the severity of the effect in affected individuals varies with the dose, and for which a threshold usually exists.

effective dose (E): The sum over specified tissues of the products of the equivalent dose in a tissue (T) and the weighting factor for that tissue (w_T), $i.e.$, $E = \Sigma \, w_T H_T$.

equivalent dose (H_T): A quantity used for radiation-protection purposes that takes into account the different probability of effects which occur with the same absorbed dose delivered by radiations with different w_R values. It is defined as the product of the averaged absorbed dose in a specified organ or tissue (D_T) and the radiation weighting factor (w_R). The unit of equivalent dose is joules per kilogram (J kg^{-1}) and its special name is the sievert (Sv).

gray (Gy): The special name for the unit of absorbed dose, kerma and specific energy imparted, 1 Gy = 1 J kg^{-1}.

high dose and high-dose rate: High doses are those levels of doses where many of the biological endpoints for low-LET radiation depart from linearity, *i.e.*, about 200 mSv. High dose rate is defined as a dose rate above which recovery and repair are unable to ameliorate the radiation damage. The dose and dose-rate effectiveness factor is assumed to apply whenever the absorbed dose is less than 200 mSv and the dose rate is less than 100 mSv h^{-1}.

Negligible Individual Dose (NID): A level of effective dose that can be dismissed. The NID is 0.01 mSv y^{-1}.

optimization: This has the same meaning as ALARA.

organ or tissue weighting factor (w_T): A factor that indicates the ratio of the risk of stochastic effects attributable to irradiation of a given organ or tissue (T) to the total risk when the whole body is uniformly irradiated.

radiation weighting factor (w_R): A factor used for radiation-protection purposes that accounts for differences in biological effectiveness between different radiations. The radiation weighting factor (w_R) is independent of the tissue weighting factor (w_T).

reference level: The predetermined value of a quantity, below a limit, which triggers a specified course of action when the value, usually a dose level, is exceeded or is expected to be exceeded.

sievert (Sv): The special name for the unit of effective dose and equivalent dose, 1 Sv = 1 J kg^{-1}.

stochastic effects: Effects, the probability of which, rather than their severity, is a function of radiation dose without threshold.

Working Level (WL): That amount of potential alpha energy in a cubic meter of air that will result in the emission of 2.08×10^{-5} joules of energy.

Working Level Month (WLM): A cumulative exposure, equivalent to exposure to one working level for a working month (170 h), *i.e.*, 2×10^{-5} Jh m^{-3} $\times$ 170 h = 3.5×10^{-3} Jh m^{-3}.

References

EPA (1978). Environmental Protection Agency. "Radiation protection guidance to federal agencies for diagnostic x rays," page 4377 in Federal Register **43** (U.S. Government Printing Office, Washington).

FDA (1985a). Food and Drug Administration. *Evaluation of Radiation Exposure from Diagnostic Radiology Examinations, General Recommendations*, HHS Publication (FDA) 85-8246 (National Technical Information Service, Springfield, Virginia).

FDA (1985b). Food and Drug Administration. *Recommendations for Evaluation of Radiation Exposure from Diagnostic Radiology Examinations*, Burkhart, R.L., Gross, R.E., Jans, R.G., McCrohan, J.L., Jr., Rosenstein, M. and Reuter, F.G., Eds., HHS Publication (FDA) 85-8247 (National Technical Information Service, Springfield, Virginia).

ICRP (1977). International Commission on Radiological Protection. *Recommendations of the International Commission on Radiological Protection*, ICRP Publication 26, Annals of the ICRP **1** (3) (Pergamon Press, Elmsford, New York).

ICRP (1981). International Commission on Radiological Protection. *Limits for Inhalation of Radon Daughters by Workers*, ICRP Publication 32, Annals of the ICRP **6** (1) (Pergamon Press, Elmsford, New York).

ICRP (1983). International Commission on Radiological Protection. *Cost-Benefit Analysis in the Optimization of Radiation Protection*, ICRP Publication 37, Annals of the ICRP **10** (2/3) (Pergamon Press, Elmsford, New York).

ICRP (1984). International Commission on Radiological Protection. *Principles for Limiting Exposure of the Public to Natural Sources of Radiation, Statement from the 1983 Washington Meeting of the ICRP*, ICRP Publication 39, Annals of the ICRP **14** (1) (Pergamon Press, Elmsford, New York).

ICRP (1985). International Commission on Radiological Protection. *Quantitative Bases for Developing a Unified Index of Harm*, ICRP Publication 45, Annals of the ICRP **15** (3) (Pergamon Press, Elmsford, New York).

ICRP (1989a). International Commission on Radiological Protection. *Optimization and Decision-Making in Radiological Protection*, ICRP Publication 55, Annals of the ICRP **20** (1) (Pergamon Press, Elmsford, New York).

61

ICRP (1989b). International Commission on Radiological Protection. *Age-dependent Doses to Members of the Public from Intake of Radionuclides: Part 1*, ICRP Publication 56, Annals of the ICRP **20** (2) (Pergamon Press, Elmsford, New York).

ICRP (1991a). International Commission on Radiological Protection. *1990 Recommendations of the International Commission on Radiological Protection*, ICRP Publication 60, Annals of the ICRP **21** (1-3) (Pergamon Press, Elmsford, New York).

ICRP (1991b). International Commission on Radiological Protection. *Annual Limits on Intake of Radionuclides by Workers Based on the 1990 Recommendations*, ICRP Publication 61, Annals of the ICRP **21** (4) (Pergamon Press, Elmsford, New York).

ICRU (1986). International Commission on Radiation Units and Measurements. *The Quality Factor in Radiation Protection*, ICRU Report 40 (International Commission on Radiation Units and Measurements, Bethesda, Maryland).

LAND, C.E. and SINCLAIR, W.K. (1991). "The relative contribution of different organ sites to the total cancer mortality associated with low-dose radiation exposure," pages 31 to 57 in *Risks Associated with Ionising Radiations*, Annals of the ICRP **22** (1) (Pergamon Press, Elmsford, New York).

NAS/NRC (1980). National Academy of Sciences/National Research Council. *The Effects on Populations of Exposure to Low Levels of Ionizing Radiation: 1980*, Report of the Committee on the Biological Effects of Ionizing Radiations, BEIR III (National Academy Press, Washington).

NAS/NRC (1990). National Academy of Sciences/National Research Council. *Health Effects of Exposure to Low Levels of Ionizing Radiation*, Report of the Committee on the Biological Effects of Ionizing Radiations, BEIR V (National Academy Press, Washington).

NAS/NRC (1991). *The Children of Atomic Bomb Survivors. A Genetic Study*, Neel, J.V. and Schull, W.J., Eds. (National Academy Press, Washington).

NCRP (1954). National Council on Radiation Protection and Measurements. *Permissible Dose from External Sources of Ionizing Radiation*, NBS Handbook 59, NCRP Report No. 17 (National Council on Radiation Protection and Measurements, Bethesda, Maryland) out of print.

NCRP (1966). National Council on Radiation Protection and Measurements. *Radiation Protection in Educational Institutions*, NCRP Report No. 32 (National Council on Radiation Protection and Measurements, Bethesda, Maryland).

NCRP (1984a). National Council on Radiation Protection and Measurements. *Exposures from the Uranium Series with Emphasis on Radon and Its Daughters*, NCRP Report No. 77 (National Council on Radiation Protection and Measurements, Bethesda, Maryland).

NCRP (1984b). National Council on Radiation Protection and Measurements. *Control of Air Emissions of Radionuclides*, NCRP Statement No. 6 (National Council on Radiation Protection and Measurements, Bethesda, Maryland).

NCRP (1984c). National Council on Radiation Protection and Measurements. *Evaluation of Occupational and Environmental Exposures to Radon and Radon Daughters in the United States*, NCRP Report No. 78 (National Council on Radiation Protection and Measurements, Bethesda, Maryland).

NCRP (1987). National Council on Radiation Protection and Measurements. *Recommendations on Limits for Exposure to Ionizing Radiation*, NCRP Report No. 91 (National Council on Radiation Protection and Measurements, Bethesda, Maryland).

NCRP (1988). National Council on Radiation Protection and Measurements. *Measurement of Radon and Radon Daughters in Air*, NCRP Report No. 97 (National Council on Radiation Protection and Measurements, Bethesda, Maryland).

NCRP (1989a). National Council on Radiation Protection and Measurements. *Exposure of the U.S. Population from Occupational Radiation*, NCRP Report No. 101 (National Council on Radiation Protection and Measurements, Bethesda, Maryland).

NCRP (1989b). National Council on Radiation Protection and Measurements. *Control of Radon in Houses*, NCRP Report No. 103 (National Council on Radiation Protection and Measurements, Bethesda, Maryland).

NCRP (1990a). National Council on Radiation Protection and Measurements. *Implementation of the Principle of As Low As Reasonably Achievable (ALARA) for Medical and Dental Personnel*, NCRP Report No. 107 (National Council on Radiation Protection and Measurements, Bethesda, Maryland).

NCRP (1990b). National Council on Radiation Protection and Measurements. *The Relative Biological Effectiveness of Radiations of Different Quality*, NCRP Report No. 104 (National Council on Radiation Protection and Measurements, Bethesda, Maryland).

NCRP (1991). National Council on Radiation Protection and Measurements. *Radon Exposure of the U.S. Population — Status of the Problem*, NCRP Commentary No. 6 (National Council on Radiation Protection and Measurements, Bethesda, Maryland).

NCRP (1992). National Council on Radiation Protection and Measurements. *Maintaining Radiation Protection Records*, NCRP Report No. 114 (National Council on Radiation Protection and Measurements, Bethesda, Maryland).

NCRP (1993a). National Council on Radiation Protection and Measurements. *Evaluation of Risk Estimates for Radiation Protection Purposes*, NCRP Report No. 115 (National Council on Radiation Protection and Measurements, Bethesda, Maryland) in press.

NCRP (1993b). National Council on Radiation Protection and Measurements. *Uncertainty in NCRP Screening Models: Atmospheric Transport, Deposition and Uptake by Humans*, NCRP Commentary No. 8 (National Council on Radiation Protection and Measurements, Bethesda, Maryland) in press.

NSC (1977). National Safety Council. *Accident Facts*, 1976 ed. (National Safety Council, Chicago).

NSC (1992). National Safety Council. *Accident Facts*, 1992 ed. (National Safety Council, Chicago).

OTAKE, M. and SCHULL, W.J. (1984). "*In utero* exposure to A-bomb radiation and mental retardation: A reassessment," Brit. J. Radiol. 57, 409-414.

OTAKE, M., SCHULL, W.J., FUJIKOSHI, Y. and YOSHIMARU, H. (1988). *Effect on School Performance of Prenatal Exposure to Ionizing Radiation in Hiroshima*, RERF Technical Report 2-88 (Radiation Effects Research Foundation, Hiroshima).

POCHIN, E.E. (1981). "A perspective on risk," in *Health and Risk Analysis*, Proceedings of the Third Oak Ridge National Laboratory's Life Sciences Symposium (Franklin Institute Press, Philadelphia).

PRESTON, D.L. and PIERCE, D.A. (1988). "The effect of change in dosimetry on cancer mortality risk estimates in the atomic bomb survivors," Radiat. Res. 114, 437-466.

SCHULL, W.J., OTAKE, M. and YOSHIMARU, H. (1988). *Effect on Intelligence Test Score of Prenatal Exposure to Ionizing Radiation in Hiroshima and Nagasaki; A Comparison of the Old and New Dosimetry Systems*, RERF Revised Technical Report 3-88 (Radiation Effects Research Foundation, Hiroshima).

SHIMIZU, Y., KATO, H., SCHULL, W.J., PRESTON, D.L., FUJITU, S. and PIERCE, D.A. (1987). *Life Span Study Report 11, Part I: Comparison of Risk Coefficients for Site Specific Cancer Mortality Based on the DS86 and T65DR Shielded Kerma and Organ Doses*, RERF Technical Report 87 (Radiation Effects Research Foundation, Hiroshima).

SHIMIZU, Y., KATO, H. and SCHULL, W.J. (1990). "Studies of the mortality of A-bomb survivors. 9. Mortality, 1950-1985; Part 2: Cancer mortality based on the recently revised doses (DS86)," Radiat. Res. **121**, 120-141.

SINCLAIR, W.K. (1985). "The implications of risk information for the NCRP program," pages 223 to 237 in *Proceedings of the 20th Annual Meeting of the National Council on Radiation Protection and Measurements*, NCRP Proceedings No. 6 (National Council on Radiation Protection and Measurements, Bethesda, Maryland).

THOMPSON, D., MABUCHI, K., RON, E., SODA, M., TOKUNAGA, M., OCHIKUBO, S., SUGIMOTO, S., IKEDA, T., TERASAKI, M., IZUMI, S. and PRESTON, D. (1992). *Solid Tumor Incidence in Atomic Bomb Survivors, 1958-87*, RERF Technical Report 5-92 (Radiation Effects Research Foundation, Hiroshima).

UNSCEAR (1977). United Nations Scientific Committee on the Effects of Atomic Radiation. *Sources and Effects of Ionizing Radiation*, Report to the General Assembly with Annexes (United Nations Publications, New York).

UNSCEAR (1982). United Nations Scientific Committee on the Effects of Atomic Radiation. *Ionizing Radiation: Sources and Biological Effects*, Report to the General Assembly with Annexes (United Nations Publications, New York).

UNSCEAR (1986). United Nations Scientific Committee on the Effects of Atomic Radiation. *Genetic and Somatic Effects of Ionizing Radiation*, Report to the General Assembly with Annexes (United Nations Publications, New York).

UNSCEAR (1988). United Nations Scientific Committee on the Effects of Atomic Radiation. *Sources, Effects and Risks of Ionizing Radiation*, Report to the General Assembly with Annexes (United Nations Publications, New York).

WILSON, R. and CROUCH, E.A.C. (1982). *Risk/Benefit Analysis* (Ballinger Publishing Company, Cambridge, Massachusetts).

YOSIMOTO, Y., SODA, M. and MABUCHI, K. (1992). "Health risks of atomic bomb survivors: The experience of those exposed *in utero* and early childhood," pages 80 to 85 in *Proceedings of the International Conference on Radiation Effects and Protection* (Japanese Atomic Energy Research Institute, Tokyo).

The NCRP

The National Council on Radiation Protection and Measurements is a nonprofit corporation chartered by Congress in 1964 to:

1. Collect, analyze, develop and disseminate in the public interest information and recommendations about (a) protection against radiation and (b) radiation measurements, quantities and units, particularly those concerned with radiation protection.
2. Provide a means by which organizations concerned with the scientific and related aspects of radiation protection and of radiation quantities, units and measurements may cooperate for effective utilization of their combined resources, and to stimulate the work of such organizations.
3. Develop basic concepts about radiation quantities, units and measurements, about the application of these concepts, and about radiation protection.
4. Cooperate with the International Commission on Radiological Protection, the International Commission on Radiation Units and Measurements, and other national and international organizations, governmental and private, concerned with radiation quantities, units and measurements and with radiation protection.

The Council is the successor to the unincorporated association of scientists known as the National Committee on Radiation Protection and Measurements and was formed to carry on the work begun by the Committee.

The Council is made up of the members and the participants who serve on the scientific committees of the Council. The Council members who are selected solely on the basis of their scientific expertise are drawn from public and private universities, medical centers, national and private laboratories and industry. The scientific committees, composed of experts having detailed knowledge and competence in the particular area of the committee's interest, draft proposed recommendations. These are then submitted to the full membership of the Council for careful review and approval before being published.

The following comprise the current officers and membership of the Council:

Officers

President	CHARLES B. MEINHOLD
Vice President	S. JAMES ADELSTEIN
Secretary and Treasurer	W. ROGER NEY
Assistant Secretary	CARL D. HOBELMAN
Assistant Treasurer	JAMES F. BERG

Members

SEYMOUR ABRAHAMSON	JACOB I. FABRIKANT	FRED A. METTLER, JR.
S. JAMES ADELSTEIN	R.J. MICHAEL FRY	WILLIAM A. MILLS
PETER R. ALMOND	THOMAS F. GESELL	DADE W. MOELLER
LYNN R. ANSPAUGH	ETHEL S. GILBERT	GILBERT S. OMENN
JOHN A. AUXIER	ROBERT A. GOEPP	LESTER J. PETERS
HAROLD L. BECK	JOEL E. GRAY	JOHN W. POSTON, SR.
MICHAEL A. BENDER	ARTHUR W. GUY	ANDREW K. POZNANSKI
B. GORDON BLAYLOCK	ERIC J. HALL	CHESTER R. RICHMOND
BRUCE B. BOECKER	NAOMI H. HARLEY	GENEVIEVE S. ROESSLER
JOHN D. BOICE, JR.	WILLIAM R. HENDEE	MARVIN ROSENSTEIN
ROBERT L. BRENT	DAVID G. HOEL	LAWRENCE N. ROTHENBERG
A. BERTRAND BRILL	F. OWEN HOFFMAN	MICHAEL T. RYAN
ANTONE L. BROOKS	DONALD G. JACOBS	KEITH J. SCHIAGER
PAUL L. CARSON	A. EVERETTE JAMES, JR.	ROBERT A. SCHLENKER
MELVIN W. CARTER	JOHN R. JOHNSON	ROY E. SHORE
JAMES E. CLEAVER	BERND KAHN	DAVID H. SLINEY
FRED T. CROSS	KENNETH R. KASE	PAUL SLOVIC
STANLEY B. CURTIS	HAROLD L. KUNDEL	RICHARD A. TELL
GAIL DE PLANQUE	CHARLES E. LAND	WILLIAM L. TEMPLETON
SARAH DONALDSON	JOHN B. LITTLE	THOMAS S. TENFORDE
PATRICIA W. DURBIN	HARRY R. MAXON	RALPH H. THOMAS
CARL H. DURNEY	ROGER O. MCCLELLAN	JOHN E. TILL
KEITH F. ECKERMAN	BARBARA J. MCNEIL	ROBERT L. ULLRICH
CHARLES M. EISENHAUER	CHARLES B. MEINHOLD	F. WARD WHICKER
THOMAS S. ELY	MORTIMER L. MENDELSOHN	MARVIN C. ZISKIN

Currently, the following subgroups are actively engaged in formulating recommendations:

SC 1 Basic Radiation Protection Criteria
    SC 1-2 The Assessment of Risk for Radiation Protection Purposes
    SC 1-3 Collective Dose
    SC 1-4 Extrapolation of Risk from Non-human Experimental Systems to Man
SC 9 Structural Shielding Design and Evaluation for Medical Use of X Rays and Gamma Rays of Energies Up to 10 MeV
SC 16 X-Ray Protection in Dental Offices
SC 46 Operational Radiation Safety
    SC 46-2 Uranium Mining and Milling-Radiation Safety Programs
    SC 46-8 Radiation Protection Design Guidelines for Particle Accelerator Facilities
    SC 46-9 ALARA at Nuclear Plants
    SC 46-10 Assessment of Occupational Doses from Internal Emitters
    SC 46-11 Radiation Protection During Special Medical Procedures
    SC 46-12 Determination of the Effective Dose Equivalent (and Effective Dose) to Workers for External Exposure to Low-LET Radiation
SC 57 Dosimetry and Metabolism of Radionuclides
    SC 57-2 Respiratory Tract Model
    SC 57-9 Lung Cancer Risk
    SC 57-10 Liver Cancer Risk
    SC 57-14 Placental Transfer

SC 57-15 Uranium
SC 57-16 Uncertainties in the Application of Metabolic Models
SC 63 Radiation Exposure Control in a Nuclear Emergency
SC 63-1 Public Knowledge
SC 64 Radionuclides in the Environment
SC 64-6 Screening Models
SC 64-16 Uncertainties in Application of Screening Models
SC 64-17 Uncertainty in Environmental Transport in the Absence of Site Specific Data
SC 65 Quality Assurance and Accuracy in Radiation Protection Measurements
SC 66 Biological Effects and Exposure Criteria for Ultrasound
SC 69 Efficacy of Radiographic Procedures
SC 72 Radiation Protection in Mammography
SC 75 Guidance on Radiation Received in Space Activities
SC 77 Guidance on Occupational and Public Exposure Resulting from Diagnostic Nuclear Medicine Procedures
SC 83 Identification of Research Needs for Radiation Protection
SC 84 Radionuclide Contamination
SC 84-1 Contaminated Soil
SC 84-2 Decontamination and Decommissioning of Facilities
SC 85 Risk of Lung Cancer from Radon
SC 86 Hot Particles in the Eye, Ear or Lung
SC 87 Radioactive and Mixed Waste
SC 87-1 Waste Avoidance and Volume Reduction
SC 87-2 Waste Classification Based on Risk
SC 87-3 Performance Assessment
SC 88 Fluence as the Basis for a Radiation Protection System for Astronauts
SC 89 Nonionizing Electromagnetic Fields
SC 89-1 Biological Effects of Magnetic Fields
SC 89-2 Practical Guidance on the Evaluation of Human Exposure to Radiofrequency Radiation
SC 89-3 Extremely Low-Frequency Electric and Magnetic Fields
SC 90 Precautions in the Management of Patients Who have Received Therapeutic Amounts of Radionuclides
SC 91 Radiation Protection in Medicine

In recognition of its responsibility to facilitate and stimulate cooperation among organizations concerned with the scientific and related aspects of radiation protection and measurement, the Council has created a category of NCRP Collaborating Organizations. Organizations or groups of organizations that are national or international in scope and are concerned with scientific problems involving radiation quantities, units, measurements and effects, or radiation protection may be admitted to collaborating status by the Council. Collaborating Organizations provide a means by which the

NCRP can gain input into its activities from a wider segment of society. At the same time, the relationships with the Collaborating Organizations facilitate wider dissemination of information about the Council's activities, interests and concerns. Collaborating Organizations have the opportunity to comment on draft reports (at the time that these are submitted to the members of the Council). This is intended to capitalize on the fact that Collaborating Organizations are in an excellent position to both contribute to the identification of what needs to be treated in NCRP reports and to identify problems that might result from proposed recommendations. The present Collaborating Organizations with which the NCRP maintains liaison are as follows:

American Academy of Dermatology
American Association of Physicists in Medicine
American College of Medical Physics
American College of Nuclear Physicians
American College of Occupational and Environmental Medicine
American College of Radiology
American Dental Association
American Industrial Hygiene Association
American Institute of Ultrasound in Medicine
American Insurance Services Group
American Medical Association
American Nuclear Society
American Podiatric Medical Association
American Public Health Association
American Radium Society
American Roentgen Ray Society
American Society of Radiologic Technologists
American Society for Therapeutic Radiology and Oncology
Association of University Radiologists
Bioelectromagnetics Society
College of American Pathologists
Conference of Radiation Control Program Directors
Electric Power Research Institute
Federal Communications Commission
Federal Emergency Management Agency
Genetics Society of America
Health Physics Society
Institute of Nuclear Power Operations
International Brotherhood of Electrical Workers
National Aeronautics and Space Administration
National Electrical Manufacturers Association

National Institute of Standards and Technology
Nuclear Management and Resources Council
Radiation Research Society
Radiological Society of North America
Society of Nuclear Medicine
United States Air Force
United States Army
United States Department of Energy
United States Department of Housing and Urban Development
United States Department of Labor
United States Environmental Protection Agency
United States Navy
United States Nuclear Regulatory Commission
United States Public Health Services
Utility Workers Union of America

The NCRP has found its relationships with these organizations to be extremely valuable to continued progress in its program.

Another aspect of the cooperative efforts of the NCRP relates to the Special Liaison relationships established with various governmental organizations that have an interest in radiation protection and measurements. This liaison relationship provides: (1) an opportunity for participating organizations to designate an individual to provide liaison between the organization and the NCRP; (2) that the individual designated will receive copies of draft NCRP reports (at the time that these are submitted to the members of the Council) with an invitation to comment, but not vote; and (3) that new NCRP efforts might be discussed with liaison individuals as appropriate, so that they might have an opportunity to make suggestions on new studies and related matters. The following organizations participate in the Special Liaison Program:

Australian Radiation Laboratory
Commissariat a l'Energie Atomique (France)
Commission of the European Communities
Defense Nuclear Agency
Federal Emergency Management Agency
International Commission on Non-Ionizing Radiation Protection
Japan Radiation Council
National Radiological Protection Board (United Kingdom)
National Research Council (Canada)
Office of Science and Technology Policy
Office of Technology Assessment
Ultrasonics Institute (Australia)

United States Air Force
United States Coast Guard
United States Department of Health and Human Services
United States Department of Transportation
United States Nuclear Regulatory Commission

The NCRP values highly the participation of these organizations in the Special Liaison Program.

The Council also benefits significantly from the relationships established pursuant to the Corporate Sponsor's Program. The program facilitates the interchange of information and ideas and corporate sponsors provide valuable fiscal support for the Council's program. This developing program currently includes the following Corporate Sponsors:

Commonwealth Edison
Consumers Power Company
Eastman Kodak Company
EG&G Rocky Flats
Public Service Electric and Gas Company
Southern California Edison Company
Westinghouse Electric Corporation
3M

The Council's activities are made possible by the voluntary contribution of time and effort by its members and participants and the generous support of the following organizations:

Alfred P. Sloan Foundation
Alliance of American Insurers
American Academy of Dermatology
American Academy of Oral and Maxillofacial Radiology
American Association of Physicists in Medicine
American Cancer Society
American College of Medical Physics
American College of Nuclear Physicians
American College of Occupational and Environmental Medicine
American College of Radiology
American College of Radiology Foundation
American Dental Association
American Healthcare Radiology Administrators
American Industrial Hygiene Association
American Insurance Services Group
American Medical Association

American Nuclear Society
American Osteopathic College of Radiology
American Podiatric Medical Association
American Public Health Association
American Radium Society
American Roentgen Ray Society
American Society of Radiologic Technologists
American Society for Therapeutic Radiology and Oncology
American Veterinary Medical Association
American Veterinary Radiology Society
Association of University Radiologists
Battelle Memorial Institute
Canberra Industries, Inc.
Chem Nuclear Systems
Center for Devices and Radiological Health
College of American Pathologists
Committee on Interagency Radiation Research and
 Policy Coordination
Commonwealth of Pennsylvania
Defense Nuclear Agency
Edison Electric Institute
Edward Mallinckrodt, Jr. Foundation
EG&G Idaho, Inc.
Electric Power Research Institute
Federal Emergency Management Agency
Florida Institute of Phosphate Research
Genetics Society of America
Health Effects Research Foundation (Japan)
Health Physics Society
Institute of Nuclear Power Operations
James Picker Foundation
Martin Marietta Corporation
National Aeronautics and Space Administration
National Association of Photographic Manufacturers
National Cancer Institute
National Electrical Manufacturers Association
National Institute of Standards and Technology
Nuclear Management and Resources Council
Radiation Research Society
Radiological Society of North America
Richard Lounsbery Foundation
Sandia National Laboratory
Society of Nuclear Medicine
Society of Pediatric Radiology
United States Department of Energy
United States Department of Labor

United States Environmental Protection Agency
United States Navy
United States Nuclear Regulatory Commission
Victoreen, Inc.

Initial funds for publication of NCRP reports were provided by a grant from the James Picker Foundation.

The NCRP seeks to promulgate information and recommendations based on leading scientific judgement on matters of radiation protection and measurement and to foster cooperation among organizations concerned with these matters. These efforts are intended to serve the public interest and the Council welcomes comments and suggestions on its reports or activities from those interested in its work.

NCRP Publications

NCRP publications are distributed by the NCRP Publications' Office. Information on prices and how to order may be obtained by directing an inquiry to:

> NCRP Publications
> 7910 Woodmont Avenue
> Suite 800
> Bethesda, MD 20814-3095

The currently available publications are listed below.

NCRP Reports

No.   Title

8 *Control and Removal of Radioactive Contamination in Laboratories* (1951)
22 *Maximum Permissible Body Burdens and Maximum Permissible Concentrations of Radionuclides in Air and in Water for Occupational Exposure* (1959) [Includes Addendum 1 issued in August 1963]
23 *Measurement of Neutron Flux and Spectra for Physical and Biological Applications* (1960)
25 *Measurement of Absorbed Dose of Neutrons, and of Mixtures of Neutrons and Gamma Rays* (1961)
27 *Stopping Powers for Use with Cavity Chambers* (1961)
30 *Safe Handling of Radioactive Materials* (1964)
32 *Radiation Protection in Educational Institutions* (1966)
35 *Dental X-Ray Protection* (1970)
36 *Radiation Protection in Veterinary Medicine* (1970)
37 *Precautions in the Management of Patients Who Have Received Therapeutic Amounts of Radionuclides* (1970)
38 *Protection Against Neutron Radiation* (1971)

40 *Protection Against Radiation from Brachytherapy Sources* (1972)
41 *Specification of Gamma-Ray Brachytherapy Sources* (1974)
42 *Radiological Factors Affecting Decision-Making in a Nuclear Attack* (1974)
44 *Krypton-85 in the Atmosphere—Accumulation, Biological Significance, and Control Technology* (1975)
46 *Alpha-Emitting Particles in Lungs* (1975)
47 *Tritium Measurement Techniques* (1976)
49 *Structural Shielding Design and Evaluation for Medical Use of X Rays and Gamma Rays of Energies Up to 10 MeV* (1976)
50 *Environmental Radiation Measurements* (1976)
51 *Radiation Protection Design Guidelines for 0.1-100 MeV Particle Accelerator Facilities* (1977)
52 *Cesium-137 from the Environment to Man: Metabolism and Dose* (1977)
53 *Review of NCRP Radiation Dose Limit for Embryo and Fetus in Occupationally-Exposed Women* (1977)
54 *Medical Radiation Exposure of Pregnant and Potentially Pregnant Women* (1977)
55 *Protection of the Thyroid Gland in the Event of Releases of Radioiodine* (1977)
57 *Instrumentation and Monitoring Methods for Radiation Protection* (1978)
58 *A Handbook of Radioactivity Measurements Procedures*, 2nd ed. (1985)
59 *Operational Radiation Safety Program* (1978)
60 *Physical, Chemical, and Biological Properties of Radiocerium Relevant to Radiation Protection Guidelines* (1978)
61 *Radiation Safety Training Criteria for Industrial Radiography* (1978)
62 *Tritium in the Environment* (1979)
63 *Tritium and Other Radionuclide Labeled Organic Compounds Incorporated in Genetic Material* (1979)
64 *Influence of Dose and Its Distribution in Time on Dose-Response Relationships for Low-LET Radiations* (1980)
65 *Management of Persons Accidentally Contaminated with Radionuclides* (1980)

67 *Radiofrequency Electromagnetic Fields—Properties, Quantities and Units, Biophysical Interaction, and Measurements* (1981)
68 *Radiation Protection in Pediatric Radiology* (1981)
69 *Dosimetry of X-Ray and Gamma-Ray Beams for Radiation Therapy in the Energy Range 10 keV to 50 MeV* (1981)
70 *Nuclear Medicine—Factors Influencing the Choice and Use of Radionuclides in Diagnosis and Therapy* (1982)
71 *Operational Radiation Safety—Training* (1983)
72 *Radiation Protection and Measurement for Low-Voltage Neutron Generators* (1983)
73 *Protection in Nuclear Medicine and Ultrasound Diagnostic Procedures in Children* (1983)
74 *Biological Effects of Ultrasound: Mechanisms and Clinical Implications* (1983)
75 *Iodine-129: Evaluation of Releases from Nuclear Power Generation* (1983)
76 *Radiological Assessment: Predicting the Transport, Bioaccumulation, and Uptake by Man of Radionuclides Released to the Environment* (1984)
77 *Exposures from the Uranium Series with Emphasis on Radon and Its Daughters* (1984)
78 *Evaluation of Occupational and Environmental Exposures to Radon and Radon Daughters in the United States* (1984)
79 *Neutron Contamination from Medical Electron Accelerators* (1984)
80 *Induction of Thyroid Cancer by Ionizing Radiation* (1985)
81 *Carbon-14 in the Environment* (1985)
82 *SI Units in Radiation Protection and Measurements* (1985)
83 *The Experimental Basis for Absorbed-Dose Calculations in Medical Uses of Radionuclides* (1985)
84 *General Concepts for the Dosimetry of Internally Deposited Radionuclides* (1985)
85 *Mammography——A User's Guide* (1986)
86 *Biological Effects and Exposure Criteria for Radiofrequency Electromagnetic Fields* (1986)
87 *Use of Bioassay Procedures for Assessment of Internal Radionuclide Deposition* (1987)
88 *Radiation Alarms and Access Control Systems* (1986)
89 *Genetic Effects from Internally Deposited Radionuclides* (1987)

90 *Neptunium: Radiation Protection Guidelines* (1988)
92 *Public Radiation Exposure from Nuclear Power Generation in the United States* (1987)
93 *Ionizing Radiation Exposure of the Population of the United States* (1987)
94 *Exposure of the Population in the United States and Canada from Natural Background Radiation* (1987)
95 *Radiation Exposure of the U.S. Population from Consumer Products and Miscellaneous Sources* (1987)
96 *Comparative Carcinogenicity of Ionizing Radiation and Chemicals* (1989)
97 *Measurement of Radon and Radon Daughters in Air* (1988)
98 *Guidance on Radiation Received in Space Activities* (1989)
99 *Quality Assurance for Diagnostic Imaging* (1988)
100 *Exposure of the U.S. Population from Diagnostic Medical Radiation* (1989)
101 *Exposure of the U.S. Population from Occupational Radiation* (1989)
102 *Medical X-Ray, Electron Beam and Gamma-Ray Protection for Energies Up to 50 MeV (Equipment Design, Performance and Use)* (1989)
103 *Control of Radon in Houses* (1989)
104 *The Relative Biological Effectiveness of Radiations of Different Quality* (1990)
105 *Radiation Protection for Medical and Allied Health Personnel* (1989)
106 *Limit for Exposure to "Hot Particles" on the Skin* (1989)
107 *Implementation of the Principle of As Low As Reasonably Achievable (ALARA) for Medical and Dental Personnel* (1990)
108 *Conceptual Basis for Calculations of Absorbed-Dose Distributions* (1991)
109 *Effects of Ionizing Radiation on Aquatic Organisms* (1991)
110 *Some Aspects of Strontium Radiobiology* (1991)
111 *Developing Radiation Emergency Plans for Academic, Medical or Industrial Facilities* (1991)
112 *Calibration of Survey Instruments Used in Radiation Protection for the Assessment of Ionizing Radiation Fields and Radioactive Surface Contamination* (1991)

113 *Exposure Criteria for Medical Diagnostic Ultrasound: I.*
   Criteria Based on Thermal Mechanisms (1992)
114 *Maintaining Radiation Protection Records* (1992)
116 *Limitation of Exposure to Ionizing Radiation* (1993)

Binders for NCRP reports are available. Two sizes make it possible to collect into small binders the "old series" of reports (NCRP Reports Nos. 8-30) and into large binders the more recent publications (NCRP Reports Nos. 32-116). Each binder will accommodate from five to seven reports. The binders carry the identification "NCRP Reports" and come with label holders which permit the user to attach labels showing the reports contained in each binder.

The following bound sets of NCRP reports are also available:

Volume I.	NCRP Reports Nos. 8, 22
Volume II.	NCRP Reports Nos. 23, 25, 27, 30
Volume III.	NCRP Reports Nos. 32, 35, 36, 37
Volume IV.	NCRP Reports Nos. 38, 40, 41
Volume V.	NCRP Reports Nos. 42, 44, 46
Volume VI.	NCRP Reports Nos. 47, 49, 50, 51
Volume VII.	NCRP Reports Nos. 52, 53, 54, 55, 57
Volume VIII.	NCRP Report No. 58
Volume IX.	NCRP Reports Nos. 59, 60, 61, 62, 63
Volume X.	NCRP Reports Nos. 64, 65, 66, 67
Volume XI.	NCRP Reports Nos. 68, 69, 70, 71, 72
Volume XII.	NCRP Reports Nos. 73, 74, 75, 76
Volume XIII.	NCRP Reports Nos. 77, 78, 79, 80
Volume XIV.	NCRP Reports Nos. 81, 82, 83, 84, 85
Volume XV.	NCRP Reports Nos. 86, 87, 88, 89
Volume XVI.	NCRP Reports Nos. 90, 91, 92, 93
Volume XVII.	NCRP Reports Nos. 94, 95, 96, 97
Volume XVIII.	NCRP Reports Nos. 98, 99, 100
Volume XIX.	NCRP Reports Nos. 101, 102, 103, 104
Volume XX.	NCRP Reports Nos. 105, 106, 107, 108
Volume XXI.	NCRP Reports Nos. 109, 110, 111
Volume XXII.	NCRP Reports Nos. 112, 113, 114

(Titles of the individual reports contained in each volume are given above.)

NCRP Commentaries

No.   Title

1 *Krypton-85 in the Atmosphere—With Specific Reference to the Public Health Significance of the Proposed Controlled Release at Three Mile Island* (1980)
2 *Preliminary Evaluation of Criteria for the Disposal of Transuranic Contaminated Waste* (1982)
3 *Screening Techniques for Determining Compliance with Environmental Standards—Releases of Radionuclides to the Atmosphere* (1986), Revised (1989)
4 *Guidelines for the Release of Waste Water from Nuclear Facilities with Special Reference to the Public Health Significance of the Proposed Release of Treated Waste Waters at Three Mile Island* (1987)
5 *Review of the Publication, Living Without Landfills* (1989)
6 *Radon Exposure of the U.S. Population—Status of the Problem* (1991)
7 *Misadministration of Radioactive Material in Medicine —Scientific Background* (1991)

Proceedings of the Annual Meeting

No.   Title

1 *Perceptions of Risk,* Proceedings of the Fifteenth Annual Meeting held on March 14-15, 1979 (including Taylor Lecture No. 3) (1980)
3 *Critical Issues in Setting Radiation Dose Limits,* Proceedings of the Seventeenth Annual Meeting held on April 8-9, 1981 (including Taylor Lecture No. 5) (1982)
4 *Radiation Protection and New Medical Diagnostic Approaches,* Proceedings of the Eighteenth Annual Meeting held on April 6-7, 1982 (including Taylor Lecture No. 6) (1983)
5 *Environmental Radioactivity,* Proceedings of the Nineteenth Annual Meeting held on April 6-7, 1983 (including Taylor Lecture No. 7) (1983)

6 *Some Issues Important in Developing Basic Radiation Protection Recommendations*, Proceedings of the Twentieth Annual Meeting held on April 4-5, 1984 (including Taylor Lecture No. 8) (1985)

7 *Radioactive Waste*, Proceedings of the Twenty-first Annual Meeting held on April 3-4, 1985 (including Taylor Lecture No. 9) (1986)

8 *Nonionizing Electromagnetic Radiations and Ultrasound*, Proceedings of the Twenty-second Annual Meeting held on April 2-3, 1986 (including Taylor Lecture No. 10) (1988)

9 *New Dosimetry at Hiroshima and Nagasaki and Its Implications for Risk Estimates*, Proceedings of the Twenty-third Annual Meeting held on April 8-9, 1987 (including Taylor Lecture No. 11) (1988)

10 *Radon*, Proceedings of the Twenty-fourth Annual Meeting held on March 30-31, 1988 (including Taylor Lecture No. 12) (1989)

11 *Radiation Protection Today—The NCRP at Sixty Years*, Proceedings of the Twenty-fifth Annual Meeting held on April 5-6, 1989 (including Taylor Lecture No. 13) (1990)

12 *Health and Ecological Implications of Radioactively Contaminated Environments*, Proceedings of the Twenty-sixth Annual Meeting held on April 4-5, 1990 (including Taylor Lecture No. 14) (1991)

13 *Genes, Cancer and Radiation Protection*, Proceedings of the Twenty-seventh Annual Meeting held on April 3-4, 1991 (including Taylor Lecture No. 15) (1992)

14 *Radiation Protection in Medicine*, Proceedings of the Twenty-eighth Annual Meeting held on April 1-2, 1992 (including Taylor Lecture No. 16) (1993)

Lauriston S. Taylor Lectures

No.   Title

1 *The Squares of the Natural Numbers in Radiation Protection* by Herbert M. Parker (1977)

2 *Why be Quantitative about Radiation Risk Estimates?* by Sir Edward Pochin (1978)

3 *Radiation Protection—Concepts and Trade Offs* by Hymer L. Friedell (1979) [Available also in *Perceptions of Risk*, see above]

4 *From "Quantity of Radiation" and "Dose" to "Exposure" and "Absorbed Dose"—An Historical Review* by Harold O. Wyckoff (1980)

5 *How Well Can We Assess Genetic Risk? Not Very* by James F. Crow (1981) [Available also in *Critical Issues in Setting Radiation Dose Limits*, see above]

6 *Ethics, Trade-offs and Medical Radiation* by Eugene L. Saenger (1982) [Available also in *Radiation Protection and New Medical Diagnostic Approaches*, see above]

7 *The Human Environment—Past, Present and Future* by Merril Eisenbud (1983) [Available also in *Environmental Radioactivity*, see above]

8 *Limitation and Assessment in Radiation Protection* by Harald H. Rossi (1984) [Available also in *Some Issues Important in Developing Basic Radiation Protection Recommendations*, see above]

9 *Truth (and Beauty) in Radiation Measurement* by John H. Harley (1985) [Available also in *Radioactive Waste*, see above]

10 *Biological Effects of Non-ionizing Radiations: Cellular Properties and Interactions* by Herman P. Schwan (1987) [Available also in *Nonionizing Electromagnetic Radiations and Ultrasound*, see above]

11 *How to be Quantitative about Radiation Risk Estimates* by Seymour Jablon (1988) [Available also in *New Dosimetry at Hiroshima and Nagasaki and its Implications for Risk Estimates*, see above]

12 *How Safe is Safe Enough?* by Bo Lindell (1988) [Available also in *Radon*, see above]

13 *Radiobiology and Radiation Protection: The Past Century and Prospects for the Future* by Arthur C. Upton (1989) [Available also in *Radiation Protection Today*, see above]

14 *Radiation Protection and the Internal Emitter Saga* by J. Newell Stannard (1990) [Available also in *Health and Ecological Implications of Radioactively Contaminated Environments*, see above]

15 *When is a Dose Not a Dose?* by Victor P. Bond (1992) [Available also in *Genes, Cancer and Radiation Protection*, see above]

16 *Dose and Risk in Diagnostic Radiology: How Big? How Little?* by Edward W. Webster (1992) [Available also in *Radiation Protection in Medicine* see above]

Symposium Proceedings

The Control of Exposure of the Public to Ionizing Radiation in the Event of Accident or Attack, Proceedings of a Symposium held April 27-29, 1981 (1982)

NCRP Statements

No.   Title

1 "Blood Counts, Statement of the National Committee on Radiation Protection," Radiology **63**, 428 (1954)

2 "Statements on Maximum Permissible Dose from Television Receivers and Maximum Permissible Dose to the Skin of the Whole Body," Am. J. Roentgenol., Radium Ther. and Nucl. Med. **84**, 152 (1960) and Radiology **75**, 122 (1960)

3 *X-Ray Protection Standards for Home Television Receivers, Interim Statement of the National Council on Radiation Protection and Measurements* (1968)

4   *Specification of Units of Natural Uranium and Natural Thorium, Statement of the National Council on Radiation Protection and Measurements* (1973)
5   *NCRP Statement on Dose Limit for Neutrons* (1980)
6   *Control of Air Emissions of Radionuclides* (1984)
7   *The Probability That a Particular Malignancy May Have Been Caused by a Specified Irradiation* (1992)

Other Documents

The following documents of the NCRP were published outside of the NCRP Report, Commentary and Statement series:

Somatic Radiation Dose for the General Population, Report of the Ad Hoc Committee of the National Council on Radiation Protection and Measurements, 6 May 1959, Science, February 19, 1960, Vol. 131, No. 3399, pages 482-486

Dose Effect Modifying Factors In Radiation Protection, Report of Subcommittee M-4 (Relative Biological Effectiveness) of the National Council on Radiation Protection and Measurements, Report BNL 50073 (T-471) (1967) Brookhaven National Laboratory (National Technical Information Service Springfield, Virginia)

The following documents are now superseded and/or out of print:

NCRP Reports

No.   Title

1   *X-Ray Protection* (1931) [Superseded by NCRP Report No. 3]
2   *Radium Protection* (1934) [Superseded by NCRP Report No. 4]
3   *X-Ray Protection* (1936) [Superseded by NCRP Report No. 6]
4   *Radium Protection (1938)* [Superseded by NCRP Report No. 13]
5   *Safe Handling of Radioactive Luminous Compound* (1941) [Out of Print]
6   *Medical X-Ray Protection Up to Two Million Volts* (1949) [Superseded by NCRP Report No. 18]

7 *Safe Handling of Radioactive Isotopes* (1949) [Superseded by NCRP Report No. 30]

9 *Recommendations for Waste Disposal of Phosphorus-32 and Iodine-131 for Medical Users* (1951) [Out of Print]

10 *Radiological Monitoring Methods and Instruments* (1952) [Superseded by NCRP Report No. 57]

11 *Maximum Permissible Amounts of Radioisotopes in the Human Body and Maximum Permissible Concentrations in Air and Water* (1953) [Superseded by NCRP Report No. 22]

12 *Recommendations for the Disposal of Carbon-14 Wastes* (1953) [Superseded by NCRP Report No. 81]

13 *Protection Against Radiations from Radium, Cobalt-60 and Cesium-137* (1954) [Superseded by NCRP Report No. 24]

14 *Protection Against Betatron-Synchrotron Radiations Up to 100 Million Electron Volts* (1954) [Superseded by NCRP Report No. 51]

15 *Safe Handling of Cadavers Containing Radioactive Isotopes* (1953) [Superseded by NCRP Report No. 21]

16 *Radioactive-Waste Disposal in the Ocean* (1954) [Out of Print]

17 *Permissible Dose from External Sources of Ionizing Radiation* (1954) including *Maximum Permissible Exposures to Man, Addendum to National Bureau of Standards Handbook 59* (1958) [Superseded by NCRP Report No. 39]

18 *X-Ray Protection* (1955) [Superseded by NCRP Report No. 26]

19 *Regulation of Radiation Exposure by Legislative Means* (1955) [Out of Print]

20 *Protection Against Neutron Radiation Up to 30 Million Electron Volts* (1957) [Superseded by NCRP Report No. 38]

21 *Safe Handling of Bodies Containing Radioactive Isotopes* (1958) [Superseded by NCRP Report No. 37]

24 *Protection Against Radiations from Sealed Gamma Sources* (1960) [Superseded by NCRP Reports No. 33, 34 and 40]

26 *Medical X-Ray Protection Up to Three Million Volts* (1961) [Superseded by NCRP Reports No. 33, 34, 35 and 36]

28 *A Manual of Radioactivity Procedures* (1961) [Superseded by NCRP Report No. 58]

29 *Exposure to Radiation in an Emergency* (1962) [Superseded by NCRP Report No. 42]

31 *Shielding for High-Energy Electron Accelerator Installations* (1964) [Superseded by NCRP Report No. 51]

33 *Medical X-Ray and Gamma-Ray Protection for Energies up to 10 MeV—Equipment Design and Use* (1968) [Superseded by NCRP Report No. 102]

34 *Medical X-Ray and Gamma-Ray Protection for Energies Up to 10 MeV—Structural Shielding Design and Evaluation Handbook* (1970) [Superseded by NCRP Report No. 49]

39 *Basic Radiation Protection Criteria* (1971) [Superseded by NCRP Report No. 91]

43 *Review of the Current State of Radiation Protection Philosophy* (1975) [Superseded by NCRP Report No. 91]

45 *Natural Background Radiation in the United States* (1975) [Superseded by NCRP Report No. 94]

48 *Radiation Protection for Medical and Allied Health Personnel* (1976) [Superseded by NCRP Report No. 105]

56 *Radiation Exposure from Consumer Products and Miscellaneous Sources* (1977) [Superseded by NCRP Report No. 95]

58 *A Handbook of Radioactivity Measurements Procedures*, 1st ed. (1978) [Superseded by NCRP Report No. 58, 2nd ed.]

66 *Mammography* (1980) [Out of Print]

91 *Recommendations on Limits for Exposure to Ionizing Radiation* (1987) [Superseded by NCRP Report No. 116]

NCRP Proceedings

No.   Title

2 *Quantitative Risk in Standards Setting*, Proceedings of the Sixteenth Annual Meeting held on April 2-3, 1980 [Out of Print]

Index

87